D1117454

To Lori,
In memory of your sister Karen.

ALWAYS MY HERO

The Road to Hope & Healing Following My Brother's Death in Afghanistan

RENEE NICKELL

God Bless,
Renee Nickell

LIFEWISE BOOKS

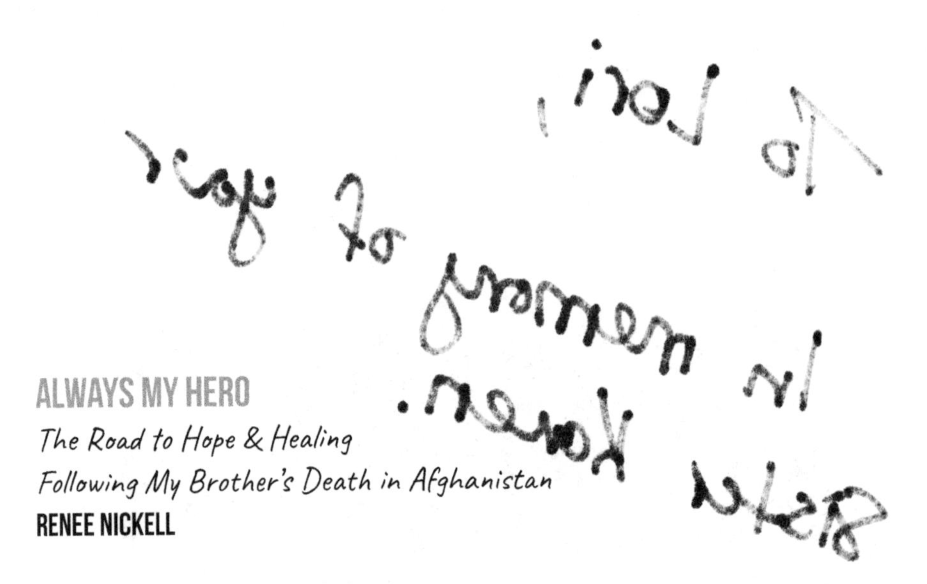

ALWAYS MY HERO

The Road to Hope & Healing Following My Brother's Death in Afghanistan

RENEE NICKELL

Disclaimer: Some names and identifying details have been changed to protect the privacy of certain individuals. While this book is a true account, the conversations gathered from personal interviews, testimonials, and memory are not necessarily word-for-word. They have been recreated from memories to give the reader a better sense of the individuals involved. Some stories have been slightly altered or information omitted to maintain anonymity. Some events may be chronologically out of order in the interest of story flow. The opinions and recollections expressed are those of the author and may or may not reflect the opinions and recollections of others portrayed in the work.

Published by:

LIFEWISE BOOKS

PO BOX 1072
Pinehurst, TX 77362
LifeWiseBooks.com

Cover Design and Interior Layout and Design | Yvonne Parks | PearCreative.ca

To contact the author:
ReneeNickell.com

ISBN (Perfect Bound): 978-1-947279-42-1
ISBN (Ebook): 978-1-947279-43-8
ISBN (Hardcover): 978-1-947279-48-3

DEDICATION

To Sam, the best brother ever and the reason this book was written…I'd rather have you.
You'll always be my hero.

1979 - Sam and Renee in Homestead, Florida.
Courtesy of Kathleen Bischoff.

SPECIAL THANKS

My husband Gentry – You have been my rock for over two decades. We've persevered through many trials together. You have seen me through some of the toughest times of my life; loving me through it all. I am forever grateful and blessed to have the most wonderful husband who loves his wife and children exceedingly.

Kylee, Leah, Ethan, and Maggie – My heart, my sweet miracles. I pray you will seek the Lord in all circumstances and persevere through every valley to reach your dreams. Thank you for loving me and supporting me as I wrote this book. You are my greatest gifts and accomplishments.

Mom – Every time I called you, crying and wanting to quit, you always told me how proud you were and gave me the pep talk to keep moving and to keep writing. You were the first one to tell me I had a "gift" and that I should pursue it. Thank you for believing in me.

Ralph P., Jason H., Blu C., Anthony C., Errol M., and Jim W. – Thank you for your contributions to my book. I couldn't have written it without you. With incredible impact and heartfelt meaning, your words captured everything that everyone loved about Sam. I know you'll always miss him. I value your friendship deeply, as it gives me a little piece of Sam to hold on to.

4th ANGLICO – Thank you for loving Sam as much as we did. Thank you for welcoming our family into yours and for continuing to honor him and keep his legacy alive. I'm still trying to keep up with all the acronyms. I'll get it one day. God bless you guys. You truly are the best of the best!

Pam and Karen – You are cousins that are more like sisters. My fondest childhood memories are with you both. Thank you for taking this journey with me. Thank you for the laughs, the tears, and the encouragement.

Amanda – The reason this book got re-started. You stirred the passion within me and got me going. From one sister to another, who knows what it's like to lose a brother, I am so very thankful for our friendship.

Cindy – You have prayed and prayed and prayed for me. You knew the ugliest parts of me yet continued to guide me and love me through it all. "Sweetie," you mean the world to me. I couldn't have made it through losing Sam and writing this book without your prayers and unconditional love.

Cyndi – You are one of the most selfless people I've ever met. Thank you for your friendship, your generosity, and encouragement. Thanks for understanding that sometimes you just need to say a bad word with no judgement. Not even country lines can keep us from being friends. I had to add what an incredible person you are for allowing me to throw you my manuscript, give you less than a 24-hour window to read it as my second set of eyes so I could meet my deadlines.

Dallas, Andrea, and Bonnie – You all have seen me through some of the worst parts of my life and yet, even with distance we can always go right where we left off. Thank you for being my heroes that day and for giving me such beautiful friendships.

Trish, Neressa, and Pastor White – You came to my rescue that fateful day and I will never forget your kindness, selflessness, generosity, and love. I often think back to what I would have done without you. You were an answer to prayer.

Jerome, Brian, & Matt – Thank you for your contributions from Sam's college years. Your stories will give my readers the comedic relief to make it through the hard stuff. No doubt you miss him and he certainly would be glad you didn't share all the "good" ones.

Jason K. – Thank you for your friendship and for reviving pieces of our childhood. You always have the best stories and I appreciate your quick wit; one of the qualities that certainly bonded you and Sam. I pray you always hold on to the memories of Sam that made your childhood so…well, memorable.

Kristen B. – I am so glad we found each other! You are such an amazing young woman and I am better for knowing you. Thank you for your contribution and for caring so much about Kylee. I am thankful for our friendship and look forward to the future.

Jaime – I have no words for the way you have impacted my life and my faith. You were an answer to prayer when I felt I had none. I am eternally grateful for our divine connection. Thank you for allowing me to share a story so sacred.

Bow Tie Media – What an incredible work you have done in helping launch this book through my website, trailer, and pictures. You guys are truly talented. Many times, you worked tirelessly on last minute jobs and I am so very grateful. You guys are going places!

Noah, Ray, Josh, Capt Oser, and the 533rd – Thank you for making that day so special without even knowing me. You guys went out of your way to extend incredible hospitality to my family. My children will never forget that day. I am very grateful for the time you generously shared to give us a glimpse into Sam's past. I have a treasure of stories I wouldn't have otherwise. Thank you so much.

Caesar – Thank you for sharing your incredible faith, your transparency, confidence, and your heartfelt words about Sam. God bless you and Annie and your family. I will hold our conversations near to my heart.

Ebony, J.D., Stephanie, and Matt – Marco…Poloooo. My peeps. The ones who saw me through from the first conception of the book sitting around the kitchen table, throwing around the idea of maybe making it happen. You guys are amazing and you've helped my faith grow stronger. Who knew what great friends we'd become? God did. J.D., I know you were looking forward to a foreword, but I'll find a place for you in the screenplay.

Bonnie and TAPS – Thank you for saving my life, not only physically, but also in the grief journey that is messy and complicated. Because of you, I've made life-long friends who just "get it." You truly care about people and that is evident in everything you do.

Charity and LifeWise Books Publishing – I stumbled upon you through circumstance, and it ended up being the greatest blessing in this writing and publishing process. Your wisdom, honesty, patience, encouragement, humor, prayers, and belief in me kept me pushing through the really tough spots. Thanks for talking me through the "unpublished" and helping me rip off the Band-Aid.

Yvonne – The cover is beautiful. When Charity told me to just "trust you", I didn't think it would be possible to create something better than what I pictured, yet you did. Thank you for creating a cover any Marine would be proud to read in public.

Kristina – Your editing, encouragement, and kindness has really made this book the best it could possibly be. You've help me make my dream a reality. I wanted something Sam would be proud of and you helped me to stay true to myself and delete the unnecessary. Thank you so much for everything!

C & N – May you always know how much you are loved, missed, and prayed for by your family. Perhaps one day when you are grown, you will learn about the stories not shared in this book; and just when you think your dad couldn't get any more honorable, you will one day learn the magnitude of his honor. I pray you take this same honor into your life with you.

TABLE OF CONTENTS

INTRODUCTION

I find no other appropriate time to begin to reflect on one's life than the age of 40. It's when the youth of our past seems "like yesterday," yet we wonder where all the years have gone and what the future will look like. Shortly after my 40th birthday, I was driving with my dear friend, Amanda, to a women's conference. She asked me, "When are you going to write that book?" It was a question I never expected, yet something I needed to hear. *When will I?*

The term "Gold Star" describes a family member who lost a loved one in the military. As a Gold Star sister, myself, I had a story to tell. I spoke to Gold Star sibling after sibling. We all had a very real pain in common–we all felt forgotten in our grief. Siblings spend more time with each other in the span of their adolescence and young adulthood than with their parents or later, with their spouses. This is not to negate the pain of the parent or the spouse, but rather to highlight that siblings are often required to suppress their pain to support the grieving process of others. I learned from a grief counselor that most siblings do not even begin the grieving process until two years after the actual event.

There are complications that happen within a family in relation to the trauma of a military death. People become un-relatable, relationships change, some families…many families, are torn apart. Parents can't cope. Spouses can't cope. Siblings can't cope. No one is able to be there for the others the way they are expected to be. I can't tell you how many Gold Star families I've met that are not able to have a meaningful relationship with their deceased child's offspring. There are broken families and broken relationships. It's all a terrible tragedy, and I was no exception.

I've discovered through my life, there is no time like the present. I could make all types of excuses. I'm a stay at home mom. I homeschool four children who need me. I don't want to expose my downfalls, my insecurities or my failures. I could choose many other excuses for not finding time to tell my story…to tell my brother's story.

Major Samuel Griffith, affectionately known as "Sam," has a story worth sharing. We both do. I decided in one moment I was going to write this book. I did not care who the target audience was going to be because I wanted everyone to read this powerful story and be changed by the person my brother was. I wanted other siblings to feel validated in their grief, and I wanted to bring some healing to other families, even if not my own. That was enough motivation right there.

As I began to reflect, I discovered much more than I had previously imagined. I am not a seasoned writer and I questioned whether I could, in fact, write a book. I investigated ghost-writing. I felt someone else could do the job better than I. Once God started closing the door to that option, I realized, no one could tell my story better than me. No one could capture my brother the way I would. This meant I had to do all the hard work, the time management, the interviews, and the research. I had to dig deep inside myself, regardless of how painful, and recall the memories, good and bad…but mostly good.

I've questioned what I should tell and what I should withhold. I debated with myself about who to mention and who to leave out. So many people made an impact on both my brother's life and my own. I wanted to give recognition where recognition was due. My only regret has been my reliance on others from our past to help me remember things I had long forgotten. In this painful process, I also began learning new things I'd never known about my brother. Many of these stories were very impactful on my life.

I traveled over 3500 miles to collect the pieces of the puzzle to make this book possible. I began interviewing people I had never met before. But they knew Sam, loved him and wanted to share this journey with me. I used to think I would be a private investigator because of my knack for researching, investigating, and seeking truth regardless of the implications. It is one of my obsessive faults. I wish it was directed more toward keeping my floors clean and my closet organized. It's not. I will choose to believe it is a God-given blessing.

I love to discover new things. Sometimes that means discovering things are less than colorful. In this instance though, I discovered beautiful things about my brother I never knew. I discovered some painful parts of his story where he confided in only his deepest, closest, most trusted confidants. I held these pieces close, yet at the same time was sad he never came to me with them. Knowing the kind of man he was, I can understand why. It was like they were lost stories in the vault of the past. This has been a healing journey not only for me, but for many others who relived their memories to help me share this story.

In these pages are stories describing who Sam was in this life. As I began to write, I hit the "delete" button many, many times before I decided the direction I wanted and needed this book to go. I discovered I was far more capable of capturing who Sam was, to me and to others, than I first believed of myself. That was the goal.

This book evolved from being a memoir about my life into this beautiful gift to all those who knew and loved Sam. If you didn't know him, you will feel like you do after reading this. I'm sure you'll regret never meeting him. He captured the essence of what it was to genuinely care about others, with no notoriety or recognition. That made him so easy to love.

During this process, I had to also discover who I was. I had to examine my own heart and my own motives. This is why I believe God called me to change the entire book after the first 30,000 words were written. It didn't feel quite right and I was grasping at straws. I wanted my story told, but I also wanted to include the legacy of my brother and our sibling bond. That is when I began to pray. I stopped writing for a couple months to reflect and investigate some other information. Through divine intervention, God brought additional people into my life who helped guide my way.

I knew I did not want a "tell all" book, although I believed it would probably sell better. A friend of mine told me, "If they didn't want their behavior talked about, then they should have behaved better." While that may be true, I did not want to leave readers feeling they were snooping into the dark pain of our past while I called people out for their bad behavior. I also felt that doing so was counterproductive to the healing process…for all parties. I certainly would never want to bring dishonor to my brother, regardless of the hurts we both had to overcome. Some challenges I didn't learn about until after this book writing process began.

This journey forced me to retrace my steps as a daughter, as a wife, as a friend, as a sister, and as a mother. Most importantly, once published, my words are forever spoken; and I want my own children to remember the most important things in life–to always tell the truth and always forgive. Telling the truth can be quite painful, but not as painful as living a lie.

Truth brings freedom. Truth may also divide relationships. I have always taught my kids the same thing my father taught me: "Always choose the harder right than the easier wrong." While this life motto may not have always played out well in my life, regardless of the poor eloquence of my words of the past, or the tactlessness I displayed at times; I was always willing to sacrifice relationship over truth. I can look myself in the mirror with no regret. Some may see this as a downfall, but this is where God taught me unconditional forgiveness.

The hardest action to take towards another person is forgiveness. My children watch me forgive time and time again, not only through my own rejection, but theirs as well. I say with upmost honestly, that forgiveness is not easy. A mother's deepest desire is to protect her children, or at least it should be. I could not protect them from the actions of others against them. These innocent little children, with hearts of pure gold, had to learn some bitter, hard life lessons through my grief-lessons most adults never learn. I've sat beside them and cried with them as they prayed blessings over those who rejected them and hurt them. I lived in darkness in some parts of my past and have had to rely on God to help me find my way out. Sometimes, His light was the only light directing me. That being said, this journey wasn't just my journey as a sister, but as so many other things. I am as authentic as I could possibly be without causing already-estranged family members more pain by airing dirty laundry.

In certain places, I felt it wise to cover the identity of some who may not deserve such protection, but my mother always taught me that none of us deserve anything, and she is right. It is only through Christ Jesus that our sins are covered. Times when I did not want to forgive or I did not want to even try to make things right, she was my voice of reason. "Do it anyway," she'd say. "It is not about them. It's about Jesus and it's about releasing them. It's about displaying forgiveness to those who don't deserve it." She was right. The more I chose forgiveness, the greater the freedom I experienced. Finally, I came to the point where I was able to write this book in love.

This is a story about a girl who once had a big brother. This big brother impacted her life more than any other person in this world.

No doubt, I am not the only person who feels this way about Sam. This book would not have been possible without the stories shared by so many of Sam's friends. I cannot take credit for this story. I may have written the words, but it is truly the sharing of so many that made this possible. For that I am forever grateful. To this I say, read on. Share in the laughter and the tears. Walk away from this a better person. Perhaps we all need a little bit more of Sam, and a whole lot more of Jesus.

1982 - Sam and Renee.
Courtesy of Kathleen Bischoff.

CHAPTER 1

OUR CHILDHOOD

The River

I opened the front door. There she was. A robust woman, with short, wavy, strawberry blond hair. Somewhere along the line in my short little life of five or six years, she had placed the fear of God in me. I was scared to death when I saw her because she and I both knew I had just been somewhere I was specifically told not to go. My grandmother, whom the family lovingly referred to as Nanny, cared for my brother and I for the majority of our lives from the ages of three to ten. My mother left when I was three, although she remained a part of our lives, and my father worked full-time. We lived in a small town in a mountainous region of Pennsylvania and shared a winding driveway with my Nanny and Papa.

My life growing up was just as complicated as anyone else's. Some memories from my childhood remain so vivid in my mind.

2015 - Nanny and Papa's house in Dauphin, PA.
Courtesy of Google Maps.

Some, I wish I could forget, while others are still a pleasant taste of the past. Somehow, when you lose someone as close as a sibling, you remember those sweet memories and, somehow the bad ones fade away.

Since the age of four, I roamed free with my brother, Sam. He was just two years older than I. Born in the 70's era, we weren't concerned with the things we are today. My big brother was my protector. I didn't need much more than that. Somehow, though, we always managed to be up to something we weren't supposed to do and always ended up getting caught. Well, not always, but every secret can't be shared or I won't have anything to keep for myself.

I approached the solid brown front door. It was the door to the neighbor's small, little yellow house, with vinyl siding and a wood deck stretching around the back side. The peep hole was too high for me to look out of. If we weren't at home, we'd surely be found there. When I say "neighbor", I mean the people who lived several acres over, as my grandparents owned many, many acres of land. They didn't have pop-up subdivisions in our neck of the woods.

Carla, a stay at home mom, and Michael, an overweight Pizza Hut manager, had two children about our age and we were fast friends, most likely because we were the only kids to play with that were close by. I wondered how Michael held his pants up with a belly that lapped over so far you could often see his button-down shirts about to burst open.

I don't know who came up with the idea, but we decided to go down to the forbidden railroad tracks. Just beyond those tracks was the Susquehanna River. Forbidden. Railroad tracks. River. They lured us, calling our names. It hadn't rained in quite a while, which meant the river would be low and we needn't worry about getting washed away.

Those creepy, old railroad tracks also ran underneath the highway. At five years old, I didn't really comprehend danger. Besides, I was with my big brother, Sam. He would protect me from anything and was certainly the only one who could punch my arm and make me cry. I still adored him and trusted him with my whole heart. I did my best to also annoy him. I imagine he wasn't as excited as I was that I always tagged along, but I learned to be as adventurous as he was.

We continued past the railroad tracks with buckets in hand and knew exactly where we were going. We could smell the musk of the stagnant water just beyond our sight. *Oh God, what if Nanny finds out? What was the likelihood she would check on us?* Street lights weren't on and we still had plenty of daylight left. It would be hours before my father would be home from work.

Sam, Justin, Amy, and I conquered the thick brush to the river like we were the indigenous people of the Amazon. The branches scraped our arms and faces as we pushed them aside. The excitement in our bellies from getting to play in the river, plus the fact we were doing something forbidden, made it even more exciting.

There it was. The river. We stood at the grassy banks. Though only a mile wide, to us, the river went on forever. We could see some old homes in the distance if we shaded our eyes with our hands from the sun. It might as well have been the Atlantic Ocean. We made it.

Decades later, I can still feel the clear, cold river water rushing between my toes. I can still remember steadying my bare feet, trying to stay put on the slimy, mossy rocks beneath me. I didn't want to fall and get hurt; however, the real reason was I didn't want to fall on my bottom and have clear evidence of our sin. The fear of death by drowning was non-existent. Besides, the river was low. We had our buckets and began moving rocks to catch our latest pets... crawfish.

I don't think there was an insect, animal, or crustacean that scared us. We frequently brought home black, garter snakes or box turtles to keep for a few days, until they were on the brink of death. Out of sympathy for our pets, and at the insistence of our parents, we had to let them go. My father taught us which snakes not to touch, primarily rattlesnakes. We had a collection of rattler tails and thought it was pretty spectacular to watch my dad chop their heads off with a shovel, while their bodies continued to writhe and wiggle without their head or their tails.

One of our favorite pastimes was picking blood thirsty ticks off our mixed Golden Retriever, Benson. My father referred to him as a Heinz 57 dog since we didn't know exactly what he was. I'd place the tick on a rock and with all his might, my brother would crush it with another rock and we'd watch the ticks explode after its feast on Benson. If you've never seen a full tick, you wouldn't understand our satisfaction of seeing how much blood we could splatter on the rocks behind our home.

We were little children running around the woods with snakes, black bears, and lots of other fun creepy crawling things. I'm beginning to realize how sheltered my kids are. I mean, this was only 30+ years ago, but who's counting.

We got our collection of crawfish and just knew we could make it back across the tracks to Justin and Amy's house before any adult would be the wiser. We admired our new pets and wondered where we could stash them. We needed an aluminum coffee can, or something to keep them alive for a bit, at least some old Tupperware. We didn't even think about the fact there would be a high probability someone would wonder where we got crawfish. You can't exactly stuff them in your shirt and smuggle them inside the house.

Sam and I knew we had been gone longer than we should have been. Losing track of time, we knew we had to check in. When I opened the front door, I gasped in shock as my stomach sank at the sight of my grandmother. She didn't say a word. She was just standing there with her arms crossed and a scowled look on her face. I could feel the lump in my throat getting bigger as I struggled to swallow. I thought my life was over. I knew it. You see, Nan had a way of mysteriously "knowing" things—a gift of sorts. I don't know why I felt so much fear except for the sheer fact we were busted in our tracks.

Let me tell you something about grandparents, you can still fear they will whip you even if they never have before. My Nan had never struck me. The worse she had ever done was make me eat dried, hot pepper on my orange sherbet for saying the "s" word and I don't mean "stupid". I'm pretty sure I must have mimicked Sam because he sat across from me with his peppered covered ice cream as well, with nothing but a smirk on his face.

I'll never forget her words though. She looked at me and said, "You were down at the river, weren't you?" I hesitantly nodded through tears. She replied with the worst thing you could ever hear…she said, "I'll let you tell your father." *Oh, God, please no. Not that. Please do not make me tell my father. Can't YOU tell him*? I wondered if this was something my brother and I could get away with. How soon exactly did she want us to tell our father?

What a burden to bear at five or six years old. Would she secretly tell him? What would happen to our butts if he found out? But worse, he would be disappointed in us. I can honestly say, I don't recall a single spanking from my father. Not because he didn't spank us, but because I knew he loved us. But before we knew it, discipline would soon be handed over to our "soon to be" new stepmother who we feared the worst. I never remember a spanking out of anger by our dad, albeit I'm certain there had to have been at least one.

Truly, I believe the disappointment is far worse than any strike on the bottom. The bottom is padded, the heart is not. A child wishes nothing more than to please their parents, especially their fathers. What is it about the relationship between children and their fathers? The thought of being rejected by a father can nearly destroy a person. Believe me, I know. I've contemplated this many times, struggling to reconcile my own personal relationship with God, the Father, but we'll get to that later.

My brother and I did what any kid would do when caught in a forbidden act of disobedience and tasked with "telling on ourselves". We never told my father. I, to this day, do not know if my father knows Sam and I went down to the river that day. I wondered for years if my Nanny had told him. She never asked us if we confessed to him. My guess is she extended a tremendous amount of grace, knowing we would never tell him, and understood we would never disobey her again. Lesson learned.

Many of the pleasant memories from my childhood included roaming the land around my Nanny's acreage which was filled with maple trees that would turn every shade of orange and red in the fall. I can still smell the sweet grass in the summertime when my brother and I would roll down the steep hill behind her house and then race to the top to do it all over again.

1977 - Sam and Renee in Dauphin, PA.
Courtesy of Kathleen Bischoff.

A large mound of dirt made its home in our front yard. My father built our home and his Caterpillar Bulldozer sat out front next to the ten foot tall dirt pile. Sam and Jason, our childhood friend, raced up the dirt pile and stood on top of it as if they had just conquered Rome.

Jason, six years old, had almost jet-black hair much like Sam's. They could have passed for brothers. Jason said, "What are you going to be when you grow up? I'm going to be a firefighter."

Sam said, "I'm going to be a fighter pilot."

Jason replied, "Well…I'm going to be a police officer."

Sam in return said, "I'm going to be a fighter pilot," as confident as his comment before.

Jason again said, "I'm going to be soldier."

Sam said again, "I'm going to be a fighter pilot."

Sam was quite certain of his destiny from a very young age. There was no compromise. How many children actually become what they say they will be from the time they are just out of kindergarten?

Jason had a similar personality to Sam, always making jokes at every chance. Nothing has changed decades later. It is probably why our fathers were best friends, even to this day.

Sam, Jason, and I were all born in Fort Belvoir, Virginia, while our dads were stationed there in the Army in the mid 70's. We moved to Pennsylvania in 1979 where my father and mother built a ranch home on the side of a mountain near my grandparents. It wasn't too long after the move when our parent's marriage fell apart and they separated. I was too young to know what was happening. All I knew was there was me and there was Sam, and we'd find any chance we could to find adventures. Our parents separated, but our bond as small children grew even closer.

We were best friends and worst enemies. We'd play together until we got sick of each other. I'd be the one who would usually end up crying because Sam would hold me down and grab my wrists, giving me "Indian burns". I'd try to grab the TV remote and he'd hold up his fist promising to give me a knuckle sandwich. I never knew what that meant when I was little but knew it must be bad and I didn't want one. Ten minutes later we'd be sitting in front of the TV watching Air Wolf, Knight Rider, or MacGyver. I always put up with those heroic, masculine, television programs, only because I wanted to be doing what Sam was doing…and he was bigger than me, which meant he always got dibs.

On days it would snow, Sam and I would use our sleds to first carve out the path we wanted to take down the steep, narrow, driveway. "You better make sure you dig deep enough or you'll go off course and collide with those trees," my brother would say. "Here, just let me show you so you do it right." He continued to look out for me. Large oak trees surrounded our home and lined our quarter mile driveway. We honestly did have to walk uphill in the snow to and from the bus stop. The bitter, cold froze our noses almost to the point of frostbite, but we'd run inside as quick as we could and sit with our backs pressed up against the fireplace to warm up. Then we'd run right back out again for more sledding.

How I loved my brother. There is nothing sweeter than the bond between siblings, especially siblings close in age. Any parent, who secretly observes this relationship when siblings are getting along, knows how endearing it is. The same sibling who didn't know better than to bite their wee baby sister's toes or hit her on the head with wooden alphabet blocks when she was just a few days old. Everyone *thinks* Sam was an angel, buuut I'm not so sure my parents thought that all the time.

1980 - Sam outside home in Dauphin, PA headed to school.
Courtesy of Kathleen Bischoff.

The Early Years

Sam had thick, dark brown, straight hair. He was always thin until he turned about 30 and realized he had to work a little harder to remain thin and trim. He loved food, particularly nachos. He could take an entire large round tortilla chip and maneuver his mouth around it to get the whole chip in his mouth at once. This was a classic move by Sam and everyone in the family found it humorous, regardless of the millions of times he'd done it. One time, he consumed a diet of nothing but chicken wings and salad for months. That was Sam though. I don't think he ever had enough chicken wings...or nachos...or Yuengling Beer, perhaps Jim Beam and Diet Coke as well.

Our dad and Sam had a very special and unique relationship. Dad was a gentle man towards us. He had piercing blue eyes, loved restoring Mustangs, and dreamed of one day having a father and son business. He taught Sam everything he knew about the restoration process of a car.

Dad and Sam spent countless hours in the garage or driveway fixing cars. My brother would later go on to write articles for the Mustang Monthly. He started his own online parts business and offered tutorials online. He even dedicated his website to my dad:

> *"You know, I often see dedications that are made to those who are no longer with us. Here's something different: This site is dedicated to my Dad, who taught me everything I know about Mustangs and turning wrenches. Dad's still kicking, and hopefully we have many, many more Mustang projects to do together. Thank you, Dad. I love you."* (The Coral Snake)

1982 - Sam and Dad working on the Cobra.
Courtesy of Kathleen Bischoff.

Anyone who knew Sam knew he LOVED Mustangs. It did not matter if they were old or new, but he loved the old ones the best. It was a love that was passed down from our dad long before Sam was learning his ABC's or learning to read in kindergarten. What started out as a hobby when he was a teenager, grew into something much more than that.

Sam spent years restoring his 1968 Fastback Mustang. He started his own parts company, selling them on eBay. He joined multiple online forums, offering his expertise to anyone and everyone that needed it. People who had never met Sam in person, only through forum, have had the nicest and kindest things to say about him.

Sam's pride and joy would one day not only be his sons, but his fully restored 1968 Fastback Mustang that took him years to complete. The completion truly began from the time Sam was just a little boy being taught how to turn wrenches in the garage of that old, white brick Pennsylvanian home. The ultimate goal was to partner with dad and run their own Mustang business. Dad had built a beautiful home on seven acres of land, enough for him to later also build a separate garage.

Not only did Sam love Mustangs, but he was an avid prankster or troublemaker of sorts. He was always trying to make people laugh. If he wasn't trying to get people to laugh, they'd eventually laugh at his over the top attention seeking efforts. Once in second grade, my parents got a call from the school: "Mr. Griffith, you need to come to the school and meet with the principal. Sam is here in the principal's office for his misbehavior." My mother and father both went to the school to meet with the principal. Sammy was sitting outside the office in a chair, with his head down.

"Mr. and Mrs. Griffith, Sam raised his hand today in class to ask the teacher a question." My mother and father just waited to hear why exactly they were there. "Well, Sam came to the front of the class and asked the teacher, 'What do you get when you mix a hurricane and a tornado?' His teacher did not know and so Sam said, 'A titty twister.'" That wasn't the worst part. "Sam actually attempted to twist the teacher's breast, as she slapped his hand." The principal explained that he had to paddle Sam for this behavior, and it was not acceptable to say and do such a thing to a teacher. "To make matters worse, Mr. and Mrs. Griffith, after Sam was sent to the corner, he picked his nose and, uh hem, placed his…well, his…boogers on the wall."

My mom and dad, trying not to make eye contact with each other, understood. When they walked out of that office, they looked at each other and both cracked a small bit of laughter when Sam could not see or hear. As a parent, I can relate. There are some stories you can't make up. Sam was taught a serious lesson in the treatment of women that day.

It wasn't until about 6th grade when the administrators of the school began to realize what my parents had known all along. Sam was special. Special in a way that he could take an entire jar of glue and turn it into a bounce ball, get sent to the hallway, and then use spit balls to cover the hallway walls, getting into more trouble. The straw that broke the camel's back was when Sam was caught urinating all over the bathroom rolls of toilet paper. Instead of punishing him, the school decided to test him. They determined Sam was near genius and was in desperate need of higher education. Only Sam could screw up and get promoted. Sam was so extremely smart that he was immediately accepted into another school with a Gifted and Talented program. I'm sure it was a relief not only for my parents, but for the school as well.

Sam was always doing something to make people laugh, but never out of malicious intent. He inherited that talent from my father. Sam didn't have a mean bone in his body. The two of them together were always an outlandishly fun time. They both had equally quick wits about them. Moments together were filled with my dad's infamous sayings and one liners. My dad would change his voice with every phone call as if we didn't know it was him, just to make us smile. I knew a prerequisite for my future spouse would be a great sense of humor.

1986 - Sam's baseball picture.
Courtesy of Kathleen Bischoff.

In 1988, we moved to North Carolina for my father's work. Jason's family had moved there a few years prior and my dad thought it might be a nice place to raise a family. He had remarried in 1985, so Sam and I not only inherited a new stepmother, a new baby half-sister Kelsey, but also Aunt Mabel, who liked to visit, usually when the moon was right.

Aunt Mabel was a small, frail woman, but fierce. She had a rough childhood and a mean streak in her a mile wide. Sam and I knew not to cross her. You were either in her good graces or you weren't. There was no in-between. No one ever stood up to her. They knew better.

We had a difficult time adjusting to a new state. It seemed a half a world away from our mom who still lived in Pennsylvania. She was given no say about our departure. My mom was single and working as a new nurse. She didn't have the luxury of contesting anything. Sam and I had to cope the best we could. It wasn't until I was much older that I realized how much pain the move had caused our mother, even though she never said so until I was a mother myself. I'm sure my father thought he was doing what was best at the time. Unfortunately, it hurt deeply to be so far from her. Soon after, she had nothing keeping her in Pennsylvania, and decided to move closer to her parents in south Florida.

Sam and I spent a lot of time finding things to do while my dad and stepmother worked. Many times, it meant finding nothing better to do than fight. One evening, when my parents were out, Sam and I began to terrorize each other. I would try to hit him and he'd chase me until he got in a good swing, but this time we were both angry. I chased him and trapped him in front of the dishwasher in the kitchen. My only weapon was a kitchen chair on wheels. I grabbed the chair and with everything in me, I pushed it towards Sam. As he jumped out of the way, the chair hit the dishwasher and left an incredibly large dent in the door.

We both just froze. "Do you think they'll notice?" I asked.

"Oh yeah, they'll notice," said Sam.

I knew it and he knew it. We were in for it. Neither of us were afraid of dad, but Carol, on the other hand, was extremely strict. Carol was an excellent caretaker; however, she had little patience and we feared for our lives like any kids who had just damaged a major appliance.

We survived the punishment and, of course, did not learn our lesson. On another occasion, Sam was trying to run from me and jumped over the six-foot-long glass top coffee table in the living room. When he put his foot down just a little too soon, the top of the table shattered. When my dad and Carol walked in, shattered glass was everywhere. Our antics never ended. The majority of the time, Sam and I lived outside investigating anywhere we could. Just to get out of the house, we would ride our bikes down old dirt roads for miles just to go to the store to buy some gum, candy, or a drink.

When we would run through the woods, I always stayed close to him. "Sammy, do you think Freddy Krueger is real?" I asked him as we ventured through the middle of the woods. Since he was my big brother, I believed anything he said.

"Yeah, I think he's real," he told me. My brother used to torture me with horror movies. Parents didn't filter anything back then, so a trip to the movie store meant a night of gruesome horror flicks and zero sleep for me.

"Are you serious? You really think he's real?" I said, with obvious fear in my voice.

"Yeah, I think there are supernatural things that happen, like Freddy Krueger. I wouldn't sleep if I were you," he smirked. I knew he was just trying to scare me, but the little sister in me wondered if he was telling the truth. "Race you to the house!" he said. Both of us leapt like deer through the woods in record time.

1991 - Sam (16) and Renee (14) in Fuquay-Varina, NC.
Courtesy of Kathleen Bischoff.

Sam's Teenage Antics

At sixteen, Sam became a new driver and was allowed to drive my dad's ugly old Ford Ranchero–a gray, half-car, half-truck contraption. With this new privilege, Sam drove us to youth group every Sunday night. Our church building was towards the street and the youth room, offices, and gym were to the rear of the property. There was a driveway that led from the main church through the cemetery to the back building. I told Sam that I wanted to sit in the bed of the truck. Along came his best friend, Jason, who jumped in the back with me.

Jason was like an additional big brother. He always heckled me and gave me a hard time just like Sam did. The two of them together often were a recipe for disaster. I used to get so pissed at the things they'd get away with especially with my father and stepmom on my case for any and every infraction. It was an awful double standard, but I appreciate these memories now. I always smile when I remember those carefree days.

1991 - Driveway leading to youth center behind church where Renee fell out.
Courtesy of Renee Nickell.

Jason was in the Boy Scouts with Sam and they both would later serve their country in the Armed Services, deploying in combat. Jason survived an IED explosion, getting hit with shrapnel while his friend and sergeant tragically lost his life right before his eyes. Jason would eventually serve his country again as a police officer, settling in Tennessee.

This fateful night in the church parking area, Sam stalled the car about halfway down the drive to the back office. I assumed he had stopped to make us get out and walk. Always trying to pull off some shenanigan, I hopped up on the closed tail gate. Without looking into his rear-view mirror, he restarted the car, put it in gear and stepped on the gas. I fell straight back, my head hitting the pavement. The world went black for a moment and I lay there on the ground, unable to move. Sam hadn't even realized I had fallen out.

Someone must have run to call my parents because Carol came as quickly as she could and rushed me to the Emergency Room. I remember the concern in her eyes. I went home with some pain killers and, thankfully, only a concussion and one day off from school. No doubt my brother received much greater consequences even though I don't remember his punishment. I do remember the letter he wrote me. I still have it to this day. I sat in my closet the other day and read the letter again. I remembered this moment, how kind and thoughtful he was, and how much Sam loved me.

> *Dear Renee,*
>
> *I really truly am sorry for hurting you like that. I know 'sorry' is probably a real lame word to you right now, but it's all I can say. I guess you know it won't happen again. Neither of us is stupid enough to be caught in that situation again, not that you were stupid the first time (ha, ha). I don't really know*

of anything I can say to make you feel better, but if there's anything I can do to lighten the load at school or anything like that just let me know. Just remember if anyone ever picks on you (except me, of course), tell them if they don't stop, your big brother will take them for a ride they'll never forget! Now, when we watch the Super Bowl at a party or something and a player gets a concussion, you can jump up and say, "Hey, I know what that feels like, it hurts like ____." I hope you are laughing by now. Sammy."

It was the last day of school before the summer of 1993. Sam was about to graduate in a few days. I could hardly wait to get to the bus. The "milk box bus" was what everyone called it, or "the short bus". We had a small bus because there were only a handful of us that took the hour bus ride to attend a magnet school in Raleigh, North Carolina. No doubt we were all often ridiculed, regardless of *why* we rode that bus. My brother tested into this school for his academic giftedness. I coasted along on his coattails and was grandfathered into the same school. Although I did well, I had to work much harder to succeed than he did.

As I walked down the sidewalk alongside the school, I looked up and saw my brother hanging out the window of the two-story brick building. He was dangling from the brick window sill. His legs swayed back and forth as he shimmied towards the outer window ledge.

"What the…??" I said, as I ran over. "What are you doing?" I yelled.

"Hurry, snap a picture!" he said as he swung there like a monkey from a branch. I had my camera out all ready to take last day of school photos with my friends so I quickly snapped a picture.

1993 – Where Sam hung outside the window and wrote his name in chalk.
I wish I had a picture of him hanging out of the window.
Courtesy of Renee Nickell.

During the spring in North Carolina, the school windows were always open. We had no AC or safety precautions or worry of students jumping or falling out windows. Sam's teacher turned around to face the chalkboard when he jumped out the window, grabbed hold of the side of the wall and waited. His teacher turned around, "Um, where is Sam?" The students looked around and his friends replied with a shrug of their shoulders. His teacher was quite confused at this point. I mean, the second story window was not one that anyone would just jump out of without any reaction from the students or screams as one fell. The teacher turned back toward the board. Sam climbed back up and sat back in his chair as if he had never been gone. His teacher turned around again and there was Sam sitting in his seat. Roars abounded from his classmates. He didn't forget to make his mark either. Not only did he hang outside the window, but he had taken a piece of chalk to write his name on the bricks outside – "SAM." He always left his mark on the world.

Sam was my constant, my friend, my confidant, my voice of reason, and my hero. Sam was brilliant, creative, loving and kind. I knew, no matter what, I always had Sam. He was my very best friend. I was used to standing behind him, lurking in his shadow. I felt safe there. I was comfortable there. I didn't like to be in the forefront. I liked to stand back and allow someone else to shine. Even writing this book is difficult. Sometimes God has a plan for us that makes us uncomfortable so He can mold us into what He desires us to be.

Growing up, I never fit in anywhere, except with Sam. Sam was the superstar of everything without even trying. He had the brains, the drive, the ambition and the personality. He also showed a bit of stupidity at times, but he always made every situation memorable. He was just "light" to be around...always smiling, always laughing, and always cutting up. There were occasional moments when he was quiet and reserved, taking in the world around him while deep in thought. I never knew what Sam was thinking about, but oh, that grin. I'd look at him and he'd look at me. He would just smirk and nod, one eyebrow raised, looking sideways at me as if he wanted me to ask him what he was thinking about.

He never spoke just to speak. His words were meaningful. Even if they were corrective, they were always kind. He wasn't afraid to have an opinion, yet he didn't share his opinion just to hear himself talk. He wanted his words to have impact with thought behind them. He never wished to offend another person but loved to debate a subject for a deeper level of conversation. I loved to argue a point with him. I knew I'd lose and be the one who would get mad in a sisterly way while he remained cool, calm, and collected. Afterwards, we would have a good laugh about it.

I believe that's why he became such a great leader. He was a rare breed and people gravitated to him. There was nothing annoying about him. He was himself all the time. I think people loved that about him. Despite having every right to boast of his successes, he remained humble. He always made those around him feel like he was no better than them. I admired that quality about him so much.

Moving Away

My brother didn't always want to be a Marine. Since he was a little boy, he always wanted to be a pilot. Everyone thought maybe he'd become an astronaut, his goal wasn't ever notoriety or recognition. There was no Instagram or Facebook platforms back then to boast about his plans and dreams. I do believe he wanted to make our dad proud, but I believe our father was proud of him because Sam was himself. Sam was his son. He set goals and reached them. Even if Sam had been trying to gain dad's approval, he didn't have to try very hard. Sam was just...Sam. He was a man you'd want to emulate, but never could. Footsteps you'd want to follow, but shoes you could never quite fill.

Sam was a downright unique individual with God-given personality traits. He worked hard and he attained what he set out to do. God placed a dream inside of him and he fulfilled it. He wasn't trying to be someone else. Sam was uniquely incredible, yet so humble. Sam did not even realize how great he was. That is a rare quality in a person.

During my teenage years, I don't think I would have made it without Sam. From the time I was born until I was 34 years old, he was the common denominator. He was the ONLY common denominator. He was the one person I could always count on and the one person I knew would never hurt me. This was why I knew I had to leave home at the age of 15 when he graduated from high school and left for college at Penn State University in Pennsylvania.

There was no way I was going to live in that house. My stepmother and I were constantly fighting much like many mothers and daughters do during the teenage years. My father never saw my point of view, so I knew there was no way I could stay there without Sam. While Sam and I both endured childhood hardships, Sam was thrust into his destiny, forgiving so easily, yet quiet and emotionally withdrawn at times. I, on the other hand, rebelled and completely lost who I was and any sense of meaning in my life. This is when I started to battle depression. I didn't feel like I mattered to anyone.

"I'm leaving. I'm going to live with Mom." I told Dad and Carol one day.

"What! Why? What do you mean, you're leaving?" He said angrily. I looked into his eyes and saw his heart break. Our family started to crumble. It had been broken a long time, just not acknowledged. His oldest son was leaving for college and his oldest daughter was leaving home at the same time. In my mind, I felt it was a welcome relief for them to see me go. The arguing and fighting would end. I didn't understand why he was upset. I really didn't. I didn't know why he was angry and hurt. Maybe he didn't understand why either. Maybe he didn't really want to understand. I was a fifteen-year-old girl and I knew I couldn't fight anymore. I didn't want to. I knew my father would never fight for me, so I chose to walk away. Something had to give, and I'd learned on more than one occasion to just walk away.

1993 - (left to right) grandparents Doris and John Marasco, Sam, Kathleen Bishoff (mom), and Renee at Sam's graduation in Raleigh, NC.
Courtesy of Kathleen Bischoff.

CHAPTER 2
JOURNEY TO THE MARINES

Penn State

It was the summer of 1993 when Sam started his freshman year at Penn State. Our family was ecstatic that he continued the family tradition of attending the university. I knew my dad had also spent four years in the Army, but I did not grow up around the military and I was somewhat shielded from the effects of the Gulf War. I knew it had happened, but as a child, one can feel displaced from the things happening in areas you only learn about in a geography class.

Sam also received a full NROTC scholarship. Getting into the Naval Reserve Officer Training Corps was quite an honor and a jumpstart to fulfilling his lifelong dream of becoming a fighter pilot. After he rushed with his new fraternity, Sigma Tau Gamma, Sam quickly made close friends and excelled academically as he always had in school. Sam was focused and driven, a first-born attribute that helped guide him to his dreams.

At the time, I felt no real danger for him being in the Naval ROTC program. The idea of Sam becoming a Marine hadn't even been a thought or subject of discussion. I was only sixteen at the time and I was more interested in boys than what Sam was doing in college. I mean, I was super proud of him. So much so, I insisted he be my prom date. I wanted to show off my "brother in the military" to my friends, even though he ended up babysitting me all night.

My mother, on the other hand, had a different feeling about military life. When Sam announced that he would be transitioning to the Marines, I remember her discouragement. Sam called, "Mom, I want to ask you if you would mind if I joined the Marines? I'm thinking of just going infantry instead of flying." My mom, of course, did mind, but would never hinder him from doing what he wanted to do. She wanted to ask, "You have always wanted to be a fighter pilot. What has changed?"

Instead, she replied, "Sam, do what makes you happy and I will support that." That's who my brother was and that's who my mom was. He trusted her and valued her opinion. We both knew mom would never hinder either of us from following our dreams even if she disagreed with our decision.

School was in full swing when Sam called to tell me he had a girlfriend. His first real girlfriend. I was livid. *Who was this girl and what did she want with my brother?* Sam and I were close and I needed to approve of this female before this relationship went any further, or so I thought. I had visions of punching her in the face because I was big and bad. I was prepared to let Sam know that no girl was going to replace my bond with him.

I quickly discovered my brother did not care what I thought even though I angrily voiced my opinion to him while he was at school. I was living back with my father and stepmother after my graduation when Sam decided to bring Robin home to meet the family. Robin had met Sam in the NROTC program, but would later have to resign for medical reasons.

Well, this must be serious if he's bringing her all the way to North Carolina to meet us! I thought. She walked in the door and I could see why Sam was attracted to her. She was lovely. Her smile was as warm as sunshine and her greeting to me was a hug instead of a handshake. She felt she already knew me because Sam was affectionately vocal about his family. I liked her right away…darn it! This completely wrecked any plans of a brawl breaking out. My assumption was way off base. I was wrong. My worry about her stealing my friendship away from my brother faded quickly.

We had an instant connection and very soon became close friends. I loved her. We were exactly three months apart in age. Sam was born on June 14th, I was born on August 14th, and Robin was born on November 14th. It was as if it was meant to be. The three of us formed a tight bond. Later, when I met my "to be husband", Gentry, he was instantly welcomed into our squad also. Life was great with the four of us. Even though they went back to college at Penn State, Robin and I kept in constant contact and our friendship grew into a sisterhood. She had my back and I had hers. Every time we'd all be together, there would be an abundance of laughter that would continue for the next fifteen years.

Sam was already halfway through college at Penn State when I moved back to North Carolina after high school graduation. My mom, a strong believer in setting boundaries, told me I had to leave home and I had nowhere else to go except back to North Carolina. I had become quite rebellious between the ages of sixteen and seventeen, acting out with buried hurt and anger, yet my father and Carol graciously allowed me to return to their home and helped guide me into adulthood. Carol made every effort to help me get a job and my own place to live, but I didn't care. I was too broken. I was a broken mess and wanted to do whatever I wanted.

My mom later told me it took everything in her to let me go. As I drove away, she stood there in the driveway and sobbed. She wanted to run after me, to wave her hands in the air to stop me from leaving. This wouldn't be the last time she had this feeling toward one of her children. But she couldn't stop me. She had to let me make my own mistakes. She had to let me grow up. I was a stupid teenager making stupid decisions and unsure of any significance or importance in my life, and seeking comfort in friends who…well, weren't very good friends.

I'll never forget when Sam and my dad had to come rescue me out of a horrible situation with ex-roommates and an atrocious boy I was involved with. The situation became so escalated, the cops intervened. Sam could have easily pummeled this kid if he wanted, but, thankfully, he maintained self-control. He was rational and level-headed, and he tried to find the best solution to the circumstances. That does not mean I didn't completely get it from him once we were alone. Sam berated me most of the way back to my father's house for even getting into such a situation. "What the hell were you thinking, Renee!?" Sam yelled.

"I don't know! Do you think I wanted this to happen?" I screamed back.

"You need to get your shit together and grow up! You're better than this! These are your *friends*? And what's up with the asshole? I wanted to kick his ass!" Sam was my big brother and he wanted to look out for me and guide me. I knew he wasn't being cruel. He was right. He knew how to make a point without demeaning me. He was fatherly towards me when he needed to be, yet corrective and loving in nature. My own father wasn't so gracious. It was a contrast for me but I knew Sam would be there no matter what I did…and I trusted him with my life. Unfortunately, I was so insecure and wanted attention from any boy who'd give it, even if they were terrible.

Sam had never experienced the anger I received from my father because he made better decisions than I did…most of the time, or he didn't get caught. I think there are just differences between sons and daughters. Somehow, fathers are more protective with daughters. I truly think my father's anger was the result of his worry about the decisions I made. I now understand how a parent can become angry because poor decisions can create permanent consequences and a ruined reputation of the family.

Since Sam was home one weekend, he decided to drive me up to State College for his ROTC Naval Military Ball. My cousin, Karen, had set me up with a friend and planned we would all go together. Sam drove and I was shotgun on our 9-hour trip to Pennsylvania. I saw a cup in the car's cup holder, and I was thirsty. I went to put it up to my lips and Sam yelled, "DON'T DRINK THAT!" I looked in the cup and saw the most disgusting substance. My brother had started "dipping" and I was about to consume his spit. I couldn't believe my brother was chewing tobacco after all the crap he gave me for smoking. Once he even lit a twenty-dollar bill on fire just to prove a point that I was "burning money".

1997 - Sam and Renee at Penn State Navy Ball.
Courtesy of Karen Maggi Mayes.

"I want to try it," I demanded.

He said, "No way, I'm not letting you dip."

I pleaded, "Come on, Sam, let me try it. What do you do?"

He nodded his head downward, looked at me sideways with pursed lips, handed me the can of tobacco, and told me to put just a tiny bit under my lip. By the expression on his face, I could tell he knew this was not going to go well. As he tried to coach me, partly laughing at me, I tried to keep the tobacco in my lip and I felt saliva building up in my mouth. I started to swallow. "SAM! This is so gross." My stomach began to hurt and I said, "I don't feel good, this is nasty!"

Sam shouted, "You're not supposed to swallow! Spit it out!" I spit out the tobacco into his cup. "Open up the glove box, I have napkins in there."

"Seriously, Sam, do you have enough kitchenware in your glovebox?" I asked.

"What? You never know when you're going to need a napkin and this is one of those times. So… you should be thanking me," he replied.

My father taught us two important life lessons: to always carry napkins, forks, spoons, and straws in our glove box and never leave the house without wearing clean underwear in case you're in an accident. The second one I'll never understand, because you're going to end up with soiled pants anyway.

Back at school, Sam and his best friends, Matt, Brent, and Jerome, lived in the Sig Tau house. When they lived in the house, they had different meal plans to choose from, just like living in the dorms. Some guys were on the meal plan and some were not. Those on the meal plan had the privilege of using the beverage dispensers containing lemonade and a few other non-alcoholic beverages. This was a rule that not everyone followed, and Matt and Sam were tired of the dispensers running low.

Matt and Sam both wanted to catch the guys in the house who were partaking of the beverages, that were *not* on the meal plan and *not* paying for them. They both informed those that had purchased the meal plan to not drink from the beverage dispensers until further notice. This would be for their own good. The idea was to add lemon flavored drink mix along with crushed laxative pills. Additionally, Matt and Sam removed all the toilet paper from the bathrooms and the storage room and hid them under Matt's bed.

It didn't take long before a Sig Tau brother said, "Man, I don't know what's happening to me. I must be sick. I keep having to use the bathroom." He jumped up and ran to the bathroom, with his hand behind him, holding himself, as if that was going to help him control his bowels. "Where the hell is the toilet paper?" he screamed. Matt and Sam's plan worked, and everyone knew not to drink from the dispenser if they hadn't paid up. Delivering consequences were part of Sam's leadership qualities, and it wouldn't be long before Sam was asked to be a student instructor at Fort Indiantown Gap.

1995 - Sam during physical training at Penn State ROTC.
Courtesy of Matt Ziegler.

It was the summer of 1995 and the Blue Wild Indigo was in full bloom. Its fragrant, blue-purple hue added beauty to the meadows of State College, Pennsylvania. Penn State University was just beginning the hustle and bustle of the new school year with the incoming freshman preparing to experience their new-found independence from their parents. Brian was one of these new freshmen, a new Navy ROTC recruit. "GET YOUR PANTS ON!" Sam screamed at Brian. Brian was about to endure a week of *indoctrination* at Fort Indiantown Gap. Sam was starting his junior year at Penn State in the NROTC program and was tasked as student instructor. Brian became frantic at Sam's demands, and shook as he was trying to hem his Navy dungarees he was just handed.

"MOVE FASTER!" Sam continued to yell in Brian's face.

"GET OUTSIDE…IN FORMATION! NOW!" Sam yelled.

Sam spent most of that week yelling at Brian and the rest of the incoming freshman NROTC Midshipman. What Brian did not know at the time, but would quickly learn, was that yelling and screaming at people was not Sam's strong suit. As a matter of fact, it was incredibly difficult. Sam had a smirk that could crack the toughest heart. Brian recalled that Sam's smirk was always just below the surface, ready to break through at any moment. When the smirk turned to laughter, he laughed with his entire body. His joy was so contagious, it quickly spread to everyone around him. Soon, everyone within earshot would be laughing with him.

Brian and Sam quickly became friends within a few weeks. Brian began spending time at Sig Tau and would soon pick Sam to be his big brother. Not only would a friendship form, but a brotherhood. At Brian's wedding years later, Sam did what was called, "The Celebration Dance", a Sig Tau tradition…a memory so fond, one can't help but smile.

2007 - Sam teaching others the Sig Tau Celebration Dance at Brian Conneen's wedding.
Courtesy of Brian Conneen.

Off to Flight School

It was Sam's last weekend in Quantico, Virginia before he would head to Pensacola for flight school. Jerome and Matt, and Matt's girlfriend Ashley, decided to take Sam out on the town in D.C. Jerome, now working for National Geographic as a cartographer responsible for producing maps, had special privileges they were able to take advantage of that evening. They parked in the NG parking garage and then walked to Buffalo Billiards just off DuPont Circle. A few months before, as a practical joke, a fellow Sig Tau brother decided to start a rumor that Jerome worked in a vault containing all the native photos of bare women. While there is no specific vault containing such material, the rumor stuck and became a running joke within the fraternity. The three were friends for life. Throughout the evening, they played pool together and a little shuffle board, drinking a few beers. The guys decided to do a celebratory round of shots of tequila. Ashley was the designated driver, so the three of them knew they didn't have to worry about the quantity they drank.

1998 - Sam and Matt during a night out.
Courtesy of Matt Ziegler.

As fraternity brothers go, there is no situation that is safe from a practical joke or two. Matt ordered another round of shots. Unbeknownst to Sam, Matt pulled the waiter aside and requested he serve Sam a shot of tequila, and to bring Jerome and Matt each a shot of water with a splash of cola. They repeated this a few times and it was time to call it a night. Since it was Sam's night out, Matt and Jerome paid the tab, yet Sam was quite confused as to why he was feeling more of the effects of the alcohol than his friends.

Around midnight, when they walked back to the parking garage, Jerome was required to show his ID badge. Since Jerome was the only one with a badge, the security guard refused to let Matt, Ashley, and Sam into the garage with him. Jerome began to get irritated with the guard who was obviously on a power trip and explained to the guard that they were in no shape to drive and needed Matt's girlfriend Ashley to drive them. The guard hesitantly succumbed to Jerome's plea and escorted them into the parking garage to their car.

The guard walked them into the lobby elevator and as the doors closed, Sam turned to the guard and asked, "Does this mean we don't get to see the vault?" The guard, confused, stared at him as if he was irritated by his question. He had no idea what Sam was referring to and gave no response. Once the guard was out of earshot, Jerome and Matt burst out laughing, busting Sam's hopes in his drunken state of ever seeing "the infamous vault."

After that night, it was time for Sam to prepare for and begin his new life as a Marine Officer and aviator. Most college graduates ease into their new jobs, gain some experience through internships, and might still live with mom and dad's help until they can financially support themselves. Not Sam. Sam had been planning for this moment for his entire life. He was headed for Pensacola, Florida–a naval aviator's dream.

NAS Pensacola is a small community in the panhandle of Florida. Not only does it have its beautiful beaches, but it has young, hopeful officers pursuing their dream of flying for one of the branches of the armed forces. Pensacola, best known for the Blue Angels, would now be Sam's home during his initial flight school training, where he'd have to prove himself if he had any chance of getting his first choice: The F-18. It's true, only the best of the best become fighter pilots. While other 21-year-olds were preparing resumes, Sam was preparing to begin his flight training.

Training wasn't all work and no play. All that pressure during training had to have an outlet and sometimes, the officers just needed to let loose. Sam was a young, handsome, 22-year-old Marine without the maturity level to understand potentially dangerous situations, especially when alcohol was involved at 2 a.m. Some lessons have to be learned the hard way, and unfortunately, Sam was no exception.

1998 - (left to right) Brent, Matt, Jerome, and Sam during the college years.
Courtesy of Matt Ziegler.

Sam's best friends, Jerome, Brent, and Matt would often go swimming in the Gulf after a night of drinking. The Marines and Sig Tau fraternity brothers weren't exactly light drinkers. Jerome began making his way out into the black Gulf water with only the moonlight and some balcony lights from the hotels in the distance guiding his way. He began to get tired and attempted to swim to shore. It was then that Jerome realized he wasn't making any progress. The waves began to overtake him, one after the other. Jerome was in trouble. His breathing became labored as he struggled against the current. He was trapped in a rip current, intoxicated, and out of breath.

Panic began to set in as his heart rate increased. Suddenly, Jerome remembered when caught in a rip current, you should swim parallel to the shore. What he didn't realize was that Sam was also in the same rip current. Sam also began to struggle and feared he would drown. Jerome and Sam were close together but in the darkness of the water, neither could see the other.

Sam and Jerome continued to swim out of the rip current. They hoped they would make it out alive. The waves crashed over them as if the water sought to overtake the two young men in the darkness. Finally, Jerome made it out of the current and swam frantically toward the shore. He made it to shallow water, looked over and spotted Sam just fifteen to twenty feet from him.

Sam was short of breath but was able to reach the shore. He hunched over and put his hands on his knees to help him breathe. He looked up at Jerome, "Dude, I had a real hard time getting to shore. I wasn't sure I was going to make it."

"Bro, me either. I was pretty sure that was it," Jerome replied.

"Yeah, I was pretty scared," Sam said through deep, quick breaths.

"Same," Jerome said. They sat in silence on the sand and realized what could have happened to them both. Neither one ever swam in the Gulf after a night of drinking again. Both were thankful for their quick thinking to swim parallel to the shore and grateful God spared their lives that night.

The following day, Sam called our mom. She was my brother's confidant and voice of reason when he did stupid things. It was a special relationship they had, and Sam trusted her deeply. "Mom, I have to tell you something," he said. She could almost hear his nervous smirk over the phone. He continued, "But please, please, don't tell Dad."

2004 - Sam and Jerome in Pensacola..
Courtesy of Matt Ziegler.

She knew he had done something reckless and Sam was just looking for some accountability. Our mother, a person of reason, was always available to listen to our sorry stories of unwise mistakes and we knew she would continue to love us unconditionally. That's what was and still is special about our mother's love. Even while she was corrective, she has always been a great listener, giving wisdom in love, allowing us to make mistakes, and lovingly guiding us to make better choices. We messed up at times, many times, yet she still loved us the same. Mom always felt honored in our confidence. She still holds a lot of our secrets close to her heart. Some secrets she couldn't ever do anything about. Except, she loved us through them.

CHAPTER 3
WINGS WERE MADE TO FLY

Flight School: Big Man on Campus

Everyone within the flight program at Pensacola was there for the same reason. The competition was, and still is, fierce. Not everyone who passed through those doors became a pilot. There are those who sit around and wait for something good to happen to them, but Sam sought after exactly what he wanted in life with the same passion he put forth in everything he attempted. This is where the rubber met the road. This is where Sam, at 22 years of age, was trained to fly a multi-million-dollar aircraft for our U.S. government. When you're a fighter pilot, there is one goal you keep in your mind: one day you will use your skills in combat.

You might think one just jumps into a cockpit simulator as soon as you arrive, but you would be wrong. New recruits must go through an intensive six-week ground school.

1997 - Sam as a newly commissioned 2nd Lieutenant in the USMC.
Courtesy of Kathleen Bischoff.

Aviation Pre-Flight Indoctrination prepares them to survive in water and on land, should something catastrophic happen in flight. They are taught how to survive worst-case scenarios before the aviator hopefuls even begin to fly a plane.

Sam worked tirelessly every day to prove to himself and others that he was more than deserving of his first choice. After several months of initial aviation training, Sam waited for the decision that would change his life forever. The selection process was well known. The top of the class was given the choice aircraft based on availability and the need of the branch of service. From there, selection dwindles. Sam waited for word on whether he'd be flying jets or helos (helicopters). He called home: "Mom, I'm waiting to hear if I got selected for jets. I'm not sure if I will get them. I mean...well, I'm just not sure. But... I'll take whatever they give me. I just want to fly."

Mom knew he wanted to fly jets. He was trying to convince himself he'd be okay with anything in case disappointment came. Not long after, the phone rang again. His voice was heard clearly across the room. "Mom! Mom! I GOT JETS! I GOT JETS!" His smile beamed through the phone. He had worked so hard. He was brilliant and at the top of his class. Sam qualified for the F-18. It was as if the universe had come together perfectly and Sam met his destiny...the dream he had worked for his entire life.

He was headed for Meridian, Mississippi, to learn to fly jets and he took his girlfriend, Robin, with him. Sam had some pretty decent flight training under his belt by the spring of 2000 and had started to build his flight log hours.

Three years after graduation, Sam and some of his closest fraternity brothers who were in their senior year at Penn State thought it was time to have a getaway to Key West during Spring Break. Brian and

Jerome had settled in their rental beach house waiting to hear if Sam would be able to make an appearance. Sam needed to log flight training hours and approached his flight instructor.

"Sir, I believe it would be a great opportunity for me to log some hours if I flew to Key West," Sam said, holding his bearing as seriously as he could. Anyone who knew Sam would know there was a smirk straining to burst across his face. I believe only Sam could make a request like that and have it actually be honored. His flight instructor, slightly hesitant, decided they could fly into the Boca Chica airfield located at Key West Naval Air Station, stay the night, and then fly back the next day.

1998 – Sam's first solo flight in plane 511, a T-34C Mentor, in Pensacola, FL.
Courtesy of Kathleen Bischoff.

Sam's fraternity brother, Carey, met the plane at the airfield so he could drive Sam to the rental house where his friends eagerly awaited him. Carey was living on the island and knew his way around pretty well. The two-seater T-45 Goshawk made its approach for landing.

Sam could see the crystal clear greenish blue water surrounding the island all around him. It was a pilot's dream to see such beauty in flight. The Atlantic Ocean to the left and the Gulf of Mexico to the right, I imagine any pilot would savor moments like this. Especially when they knew they may be called to fight in a war with nothing but dry, desolate, desert.

As the plane came in for landing, Carey noticed the plane begin to waggle. He knew this would be a perfect opportunity to give Sam a hard time. When Sam and the instructor came into the hanger, Carey, unable to contain himself, began to joke with Sam about his landing. An awkward silence ensued and the instructor looked at Sam, and then looked at Carey, almost as if a dog got caught red-handed in a forbidden act. The instructor admitted that it was, in fact, he that landed the plane. Laughter abounded. Sam was off the hook for that one.

For the flight back, the plan was to refuel in Daytona before returning to Meridian. Since Sam could only spend one night, he knew that he would not get to see our mom, who lived in Jupiter, Florida, just north of West Palm Beach. Again, he decided to convince his instructor that they should change their original plans. Sam wanted to refuel at Palm Beach International Airport instead of Daytona. Of course, the instructor agreed.

Sam always made every situation memorable. He lived his life to the fullest and, regardless of circumstance, he made time for all those in his life. He made each relationship count. He called Mom and asked her to meet him at the airport. Sam flew in and, with his coolest Top Gun jump out of his landed jet, greeted our mom and stepdad, Donnie. It wasn't long before Sam got his wings. The entire family drove in from all over the country. Aunts, uncles, grandparents, his siblings, friends...were all there to celebrate his accomplishment.

2000 - Sam after arriving at PBI airport exiting aircraft.
Courtesy of Kathleen Bischoff.

2000 - (left to right) Kathleen Bischoff (mom), Sam,
Donnie Bischoff (stepfather) during Sam's refuel at PBI airport.
Courtesy of Kathleen Bischoff.

Here Comes the Bride

In May of 1999, Gentry and I were getting married...and having a baby. I was six months pregnant with our first child. Gentry and I were 22 and 21, respectively, and we had no idea what we were getting ourselves into. We simply knew we were hopelessly in-love and couldn't wait to start our lives together with our little family. I can't say our families were as thrilled about the pregnancy as we were, but they were supportive in our decision to marry. Sam was a groomsman and Robin was one of my bridesmaids. It was a simpler time when everything seemed innocent without the complexities of married life or the even greater complexities of married military life. To this day, both have been the best decisions we ever made.

Just seven months later, Sam and Robin followed suit and were married in Meridian, MS. I had just had the baby. She had terrible colic which made it impossible to sleep, and even more impossible to travel out of state. I couldn't make their wedding several hundred miles away and I regret it to this day.

Once Sam earned his wings as an F-18 Aviator in Meridian, they moved to Lemoore, California. They were in their young 20's and lived at the Naval Air Station where he proudly wore the newly prestigious title of "F-18 Aviator". The population of the area located in the San Joaquin Valley was about 25,000 people. I recall this certainly wasn't their favorite place to live, but in the military community, you make friends quickly. Sam and Robin quickly assimilated into the busy life and made many friends.

2000 - Sam gets his wings in Meridian, MS.
Courtesy of Kathleen Bischoff.

Sam and Robin loved to have a good time. He lived by the motto "go big or go home". Halloween was Sam's favorite holiday and each year, he and Robin would always have a themed couples Halloween costume. One particular year while stationed in Lemoore, they could have easily won best Halloween costume. Sam dressed as Major Anthony Nelson from the 70's TV show "I Dream of Jeannie" and Robin went as Jeannie.

Robin was adorable in anything she wore, and hit the mark in her bare midriff costume. No doubt, she was the envy of her friends and the attention stealer of any man present. She was beautiful but didn't realize how beautiful she was. Sam was always so proud to have her on his arm.

Where there was partying, there was drinking, and most definitely a great story developed behind a night of drinking in the military world.

Loud music was blaring on this cool October evening. Marines and their dates filled every space of the backyard of the host pilot's home. If there was a pilot there without a date, he'd be sure to find one by the end of the night.

Sam was busy socializing, laughing, and cracking bad jokes. The smell of beer and cheap whisky filled the air. Sam realized Robin was no longer on his arm, but, as the social butterfly she was, he had no concern. There's an old southern saying that, "She could charm the dew right off a honeysuckle," and that she did. Robin was confident mingling with others and her friendly smile and infectious laugh made her popular.

Time passed and Sam noticed she wasn't anywhere to be seen. Unfortunately, during a party like this, it was common for guests to be met by a cold, bathroom floor. Concerned, Sam began to search for her. He walked down a hallway jiggling door handles. He was slightly inebriated and had forgotten where the bathroom was. He found an unlocked, cracked door. As he opened the door to one of the spare bedrooms, he found her asleep on the bed. She had a mixture of exhaustion from the day's activities and a couple drinks, and she was ready for a whisky-induced nap. He walked over to her, stared at her beauty, and knew how much he loved her. He bent down, scooped her up and knew he wanted to be the one to take care of her forever. Sam wasn't Sam without Robin. Sometimes, in those moments when we are the most vulnerable, we know who will be there for us. Sam was incredibly loyal to Robin, which was part of his nature and he never changed.

While not everyone acted the most dignified at military parties, that's not what was important. These families…these husbands…they saw and lived war. They cherished every chance they could get to enjoy a moment and they made memories. *That's* what was important.

Sam and Robin always emailed me their pictures of their Halloween costumes each year (before social media). I admired both their sense of adventure and zest to live in the moment...and so they did. What else should you do when tragedy strikes like that of 9/11?

I'll never forget that day. How could I? How could anyone who was old enough to remember? It was just around 8am and I was sitting on the edge of our bed when the phone rang. "Renee, turn on the TV, a plane just hit a tower in New York," my mom almost yelled through the phone. I turned on the TV as I hung up the phone. Horrified, Gentry and I began discussing how an accident like this could happen. All those people injured. All those lives lost.

Wide eyed, we were glued to the screen. Just then, right before our eyes, we saw it happen again...another plane flew into the second tower at exactly 8:03am. The explosion was nothing I had ever seen before. I couldn't believe it was happening. That moment, I and every other American knew...the first one could have been an accident, but the second...this was no accident. The United States was under attack.

Following suit, the Pentagon was struck at 8:37. We waited and watched with the rest of the world as screaming people ran for their lives. It was like watching a horror movie. How could this be real? At precisely 8:59, the first tower collapsed. We knew there were thousands of people who would burn to death and lose their lives prematurely. People jumped out of buildings and fell to their deaths on live TV; moms and dads who would never return home to their families. A fourth plane headed for the Capitol in Washington, D.C. for the same destructive purpose, but due to the heroic actions of its passengers, crashed in a Pennsylvania field near Shanksville. Reports came in with recordings of the heroic passengers calling their loved ones to tell them goodbye. The second tower collapsed at 9:28.

I'll never forget how Gentry looked at me that fateful morning. "Renee, we're going to war. I have to serve my country." As tears ran down my face, I raced to our mom's house just a few miles away. I pulled in quickly, leaving slight skid marks on her drive as I braked the car to a screeching halt. I ran up her porch stairs. As I flung the porch door open, she met me there. We just grabbed one another and cried. "Mom, Gentry's joining the military," I stuttered through my tears.

"Sam…I just know he'll be going too. He's been training…" she replied. That moment was repeated all over the country that day and became real to thousands of military families. Many families I know today had sons and daughters who made the heroic decision that day to serve their country and paid the ultimate sacrifice. Everyone can recall where they were and what they were doing on 9/11. It wouldn't be until later down the road that I learned that Sam's college friend, Brian, was at ground zero that day. He recently shared with me his harrowing experience on 9/11.

Brian had taken the PATH (Port Authority) train from Newark, New Jersey to the World Trade Center to attend a business meeting. He was approximately ten feet from the lobby when he heard a loud explosive crash. He thought it must have been an elevator accident. Unbeknownst to him, it was the first plane hitting the tower. Not knowing what happened, he followed others to the Trinity Church grounds. As he stood there, the second plane hit and exploded over their heads. He quickly ran to Gracie Mansion as the first building began to collapse. People were screaming and running through clouds of ash billowing from the building. When the debris cloud caught him in under a minute, he stood in a complete blackout. Once he could see, he continued to walk a few more blocks and stood in front of the federal building when the second building collapsed. Narrowly escaping with his life, he walked 70 more blocks to a hotel in which he settled for the night as he and the rest of the country mourned in shock.

Brian miraculously survived that day but will never forget his experience. He went on to marry the love of his life, and they now have two beautiful children.

I had only been married two years when I drove Gentry to the train station to ship him off to basic training. When our toddler, Kylee, wrapped her arms around his neck to kiss her daddy goodbye, I had no idea how much harder life would become for us. Not only was our future uncertain as a new family, but it was even more uncertain as a new military family who had just experienced a terrorist attack on our country. What would become of our future or the future for so many other families that shipped out that day?

Married Life in Beaufort

Sam began eighteen-month training specifically for an upcoming three-month deployment to Iraq. After Lemoore, orders came for them to move to Beaufort, South Carolina. This is where the world would be introduced to one of Sam's pride and joys: his firstborn son, Neil. Sam was assigned to the VMFA-533 squadron where he was assigned for his first deployment to Iraq, flying the F-18's. Sam was one of the new group of aviators that came into the squadron in early 2002. This was a time of preparation to prove that he was a good wingman. Aviators did what they were told and did a good job. If they obeyed their orders, they learned their mission responsibility. Sam was under a microscope getting a firehose of information as a young flyman and loving every second of it. He couldn't believe he got paid to do this job that he loved so much.

Sam was a young wingman flying locally in Beaufort and taking cross country trips as well. He frequented Pensacola, Key West, and Virginia Beach. He'd fly out to the southwest and drop a bunch of bombs in the desert of Tucson. The summer of 2002, the squadron flew to Fairbanks, Alaska and joined with the Air Force at Eielson Air Force Base. The weather was beautiful and the sun did a low 360, never setting. After flying all day, they would hit the nightlife of Fairbanks and went fishing for salmon every chance they had.

In late 2002, there were talks of going to war. The squadron was unofficially notified that they may be one of the first squadrons to support Operation Iraqi Freedom, so they adopted the mindset and began preparation. It was no longer *if* they would go, but *when*. The guys were notified that instead of Kuwait, they would be headed to Japan. They were all disappointed at their missed opportunity to go to war and put all their hard earned skills to use. Just as quickly as they got word they were going to Japan, the orders were changed back to Kuwait. Everyone got excited again!

Early 2003, they got the word that they would be leaving in February for Kuwait. All the ground forces and squadrons started heading overseas for a 90-day deployment from February to May. It was a whirlwind tour in which they got straight to work. The war officially kicked off and they began dropping a ton of ordinance…well, several tons of ordinance. The mission was cut and dried, and they headed back home in May.

2003 - Sam and his F-18 serving with the 533rd, Beaufort, SC.
Courtesy of Kathleen Bischoff.

During the summer of 2003, the guys all started to become section leads and vision leads. During the rest of that year, they earned qualifications on the airplane, which in turn would qualify them to move up to the next level of responsibility. They traveled out west a few times. It was during one of those trips to 29 Palms that Sam got the call that his first son was about to be born. Robin went into labor early. A friend rushed Sam to the airport with little time to make it from California to Beaufort, South Carolina.

Sam made it to Beaufort but didn't know if he missed the birth of his son. Another friend picked him up from the airplane and drove as quickly as they could to the hospital. Sam ran inside, frantically asking for directions to Robin's room. He ran down the hall, found her room, and swung the door open. Robin was in the last stage of labor and Neil was on his way. Sam had made it in the nick of time. Robin pushed a couple more times and Neil was born in the presence of both his mother *and* father. Tears of joy filled their eyes as Neil was lifted to Robin's chest. It was a miracle Sam made it in time to welcome Neil into the family. It was a moment so many military families sacrifice.

June, 2004, Neil was four months old and Sam received orders to head out to the West Pacific. His squadron traveled to Japan, Hawaii, Korea, Singapore, Guam, and Australia. The group was then broken into three locations for four to five weeks. Part of the squadron went to Singapore; some went to Australia while Sam and seven other flyers got the short end of the stick. They were sent to Korea as part of a command post exercise at Osan Air Base, South Korea. They lived in sparse military tents pitched in the mud. While dealing with these challenges, a destructive typhoon came through and everyone got sick.

After the five weeks of separation, the squadron came together again. Most of them boasted about their time in Singapore or Australia and how awesome it was. Sam and the seven other guys had a different story, "Yeah, you all had porcelain toilets while we slept in tents!" They weren't happy about their experience, but they made it through.

In the summer of 2005, the members of the squadron had to say, "Farewell" when they parted ways. For three years, they had shared experiences and supported each other through the tough weeks and months of training; however, no military group stays together forever. Sam and Robin were stationed in Virginia where they finally started their new life together as a family.

The RV

Recreation was a vital part of Sam and Robin's marriage. Sam was infamous for his poor joke telling ability. As a matter of fact, he could never get through a joke without laughing so hard, his whole body would shake. If it seemed that there wasn't something to laugh about, then Sam would create something. During this season, the "something" happened to be his "new" old RV. Sam and Robin bought an RV which was reminiscent of the one in the movie, "National Lampoon's Christmas Vacation." Like in the movie, they decided to take a road trip in their beautiful RV. It seemed to be the perfect opportunity for Sam to bring the family in the RV to visit Brent and Jerome who were living in Pensacola. Sam's son, Neil, was just over a year old.

"Did you hear the gorilla joke?" Sam began with a smirk. Neither had heard the joke, which thrilled Sam. He began to get very animated in his joke telling, running in place and jumping on the floor. The more the joke continued, the harder Sam laughed. By the time Sam got to the punchline, no one could understand a word he said because he was laughing so hard. His laugh was so infectious, he invoked laughter in everyone else. Brent and Jerome began rolling in laughter because they couldn't understand anything Sam was saying.

Everyone who knew Sam had one thing in common: Not a soul has ever heard the punchline to any of his jokes, even though he was known as one of the biggest joke tellers as he laughed his way through each one. I often wonder if Sam ever knew his friends and family were laughing with him and not at his jokes. Or did they think his jokes were so funny, they rolled in laughter only encouraging him to tell more? Regardless, it was an endearing trait.

The following day, they all decided to have a beach day at Gulf Islands National Seashore. The sound of seagulls could be heard above them and the cool ocean breezes brushed against their skin like a welcome relief from the hot Florida sun.

Sam and Robin brought their dog, a Hungarian Vizla, named Coby. If you don't know what a Vizla is, look it up. I can already hear you say "Oooohhh, yeah." Most of the time, when I see someone with a Vizla and identify it by its breed, people gleam from ear to ear as if they are about to hug you. It's usually followed up by, "I can't believe you know what a Vizla is!" They are great dogs. Loyal, trusting, and wonderful family members. They usually need to be owned by runners because of their exuberant amount of high energy. Fortunately, for Coby, Robin loved running...Sam, not so much.

Coby was named after the Cobra, perfectly fitting Sam's love of Mustangs. However, Brent and Jerome had a different name for Coby. They called him F.B. This was an acronym for a very special name, "Fake Balls." Nothing would be a more appropriate nickname for Sam's beloved dog. You see, Sam had Coby neutered, yet felt terrible about having to do that to his dog. Fearful that Coby would somehow lose his "manhood," Sam had testicle implants put in his dog, so he would not develop a complex. Which one of the two would have the complex? I am unsure. Nonetheless, it gave anyone who knew of Coby's cosmetic procedure the right to give Sam a hard time about it. All I could do was shake my head when he told me what he'd done. Yes, it's a real thing and yes, he really did it.

A few hours had passed at the beach and everyone was getting tired. They packed up the RV and decided it was time to head back. Sam had parallel parked the RV alongside the road, facing east. Unfortunately, they needed to go west in order to go back the way they came. Sam had not thought this through very thoroughly. So, he decided he would attempt a three-point turn, backing the RV into the beach sand.

Brent and Jerome looked at each other. "Sam, are you sure you want to do this?" Brent asked.

"Oh, we're good!" Sam replied.

Well, Sam quickly learned things weren't good. The RV got stuck in the sand. Brent, Jerome, and Sam were now digging sand out from behind the tires of the RV to "unstick" it. After many attempts, they could not get the RV out of the sand. To make matters worse, they were thick into the park and needed an extra-large tow truck to get them out. Sam was embarrassed, but they all had a great laugh. I believe he was more embarrassed that he could completely restore cars and fly multi-million-dollar fighter jets, yet he could not get his RV out of the sand.

That RV provided enough memories for our families to last a lifetime. Sam and Robin drove that RV all the way to South Florida and parked it right in my mother's front yard…in the grass. We all loved that RV. By the time it was really broken in, we all had babies and could take that thing for a spin to the mall or the grocery store with about nine of us piled in there. Gentry would scream as we exited, "It's not a clown car!"

One of those trips to Mom's was Christmas, 2006, the last Christmas we all spent together. It was a time filled with laughter, spiked egg nog, tickets to the Nutcracker Ballet (while we left all the men at home to watch children and eat nachos), and a little cottage style living room crammed full of presents from one end to the other.

2006 - Sam during our Christmas together, always smiling.
Courtesy of Kathleen Bischoff.

I don't know what became of that old RV. Years and many memories later, it sat alongside Sam and Robin's house as if it was in its own graveyard, many miles driven and well used. Funny how something can be so meaningful and bring so much laughter and joy, and eventually have its own time of departure.

CHAPTER 4

HANGING UP HIS FLIGHT SUIT

The time was coming for Sam to "qualify" to continue flying with the 533rd in Beaufort. Civilians often don't realize you must impress your superiors and go through an interview process in order to keep or change jobs in the military. The process is not just cut and dried because it's the military. Squadrons want a team player so they investigate your background and work ethic to see what you are all about.

There was one Operations Officer in particular who seemed to have it out for Sam. We'll call him "Major High Horse". This Major (OPSO) made it absolutely impossible for Sam, who was a Captain at the time, to requalify and remain an F-18 aviator. Unfortunately, there are many really amazing, talented, honest, hard-working people in the military whose optimism and work ethic rub some miserable superiors the wrong way. Major High Horse was that kind of person. Not only with Sam, but with others as well, and he was hell-bent on not allowing Sam to fly. So, Sam could not return to Beaufort to fly the F-18 because Major High Horse wouldn't allow him to qualify.

Sam came to a crossroads after this. He could have either joined the reserve unit (VMFA-112, The Cowboys) in Fort Worth, TX and fly, or he could have flown with the Navy. Sam called mom, like he'd always done when he had some major decision to make. He explained to her that he wasn't going to fly anymore and asked her what she thought about him joining the 4th ANGLICO (Air Naval Gunfire Liaison Company). Mom asked him why exactly he wanted to join *that* particular unit. Sam explained, "Mom, I just want to make a difference. I feel like maybe I can make a difference there."

Sam and Robin decided to transition to Virginia Beach after he hung up his flight suit. Because of his expertise in the air, he decided to interview with the 4th ANGLICO Reserve unit in West Palm Beach, Florida. Now, one thing about the 4th ANGLICO was pretty well known. These guys were like the cream of the crop in the Marine world. They're called the "special eyes" force. You either get invited by them or you have to try out. They don't take just anyone. There is no doubt why Sam wanted to run with these guys. They are some of the best people and they have amazing training budgets. So, if you get the chance to run with that prestigious group, it's hard to want to do anything else.

Sam interviewed with the C.O. After the interview process was over, the C.O. told Sam, "Major Griffith, you have a lot of experience and strong leadership skills. I really believe you can make a difference here." Sam called mom after the interview and told her he was offered the position. He felt what the C.O. said to him was confirmation that he was making the right decision. Now, Sam with his experience flying, qualified to be 4th ANGLICO's Forward Air Controller where he was their eyes in the sky for bombing. Sam was very good at what he did. He was very well known on the systems, pretty much a subject matter expert after his time on active duty.

It was February 2009, and Sam, Robin, Gentry and I were at my dad's house for a visit. I remember my dad tell me about Sam's upcoming deployment to Iraq. I could see the fear in his eyes. "Renee, this time…he'll be on the ground. It's very dangerous," Dad explained.

I remembered those words, yet for some reason, I wasn't afraid. I somehow felt Sam was invincible. I knew he was skilled in what he did. I knew he was brilliant and wise, and I knew he wouldn't make any mistakes over there. I brushed off what my dad said that day in the kitchen. It was as if I just knew Sam would make it home safely. Sam's deployment went off without a hitch, and he did, in fact, come home safely.

2009 - Sam as a Forward Air Controller with the 4th ANGLICO, Iraq during Operation Iraqi Freedom.

2009 - (left to right) Sam, Errol Miranda.
Matthew Tricarico, Anthony Cuesta, and George Mills.
Courtesy of Anthony Cuesta.

Over time, I noticed that Sam became quieter, often deep in thought. At times, I felt he was disconnected and withdrawn. You hear the term Post Traumatic Stress Disorder (PTSD) and brush it off because you don't really know what it is. To service members who have spent their time in war and have seen more horror than the large majority of us ever see, PTSD is very real. Did Sam have that? I don't know if he did. I never discussed it with him. Maybe there was something else on Sam's mind that distracted his focus. Now, I wished I had asked him, but Sam did not like to talk about emotions or personal feelings. He may have alluded to some things or given his opinion on matters of the heart, but it was rare to hear Sam share his personal feelings about anything. Part of me understands that. Actually, I understand that a lot. Sharing feelings was not a large part of our upbringing and I doubt he'd suddenly start at 34 years of age. This issue affected us both.

It was a cold winter Sunday night in Virginia Beach and the ground was covered in white, powdery snow. Sam made most of his phone calls to family on Sundays. Christmas was just around the corner. He had called mom to check in on her, "Hey, Mom!" Sam said.

"Hey, honey, how are you? Have you made your Christmas plans yet? I know you mentioned you weren't coming home this year."

"Oh, Mom, Christmas is just another day," he said flippantly.

"Sam! How could you say that about the birth of our Lord and Savior? Why would you say such a thing?" she said strongly.

"Well, Mom…that's just the way I feel. It's just another day to me," he replied quite matter-of-factly.

Mom did not voice her anger, but she was quite upset after she hung up, pondering why Sam would say that. Her heart was broken that he would think Christmas was just another day.

A few days later, the phone rang. It was 7 am, and Sam was calling Mom.

She wondered why he was calling her so early because it was so unlike him. She picked up the phone, "Sam? Is everything okay?"

"Mom." Sam could hardly get her name out. He was crying so hard, he had to pull his car over on the side of the road.

"Sam, is everything ok?" Mom asked. She couldn't make out what he was saying. All she could hear was his uncontrollable sobs through the phone.

"Sam…I can't understand what you're saying." She now grew quite concerned.

"Mom…I'm so sorry. I'm so sorry. I'm so sorry I said that the other night. I watched a Charlie Brown Christmas the other night. When Linus started reciting the Book of Luke, it just struck me. I realized how wrong I was. I just want to tell you how sorry I am, Mom."

Mom was taken aback. She didn't know what to say. Sam, so vulnerable in that moment, brought her to tears. "Sam, it's ok. It's ok."

Years later, Mom shared that very personal story with her Pastor right before Christmas. On a Sunday morning before Christmas, that pastor spoke of a young man and told that story to his congregation within his sermon, glancing over at my mom, as tears filled her eyes. He expressed that sometimes when we are in a place of unbelief or perhaps feeling distant from God, that God can use the smallest circumstance to help trigger a heart response from us to God.

Sam never imagined his impact on the world. That story, so intimately shared with his mom through the tears of a Marine, would one day be a part of his legacy.

Life is Good

One of the many wonderful things about having an older brother close to my age, who also had a family, was our family "vacays". All of our kids were little, so there were always temper tantrums, spankings, laughter, fights, giggles, and injuries. Family vacations were a blast! The parks were a perfect place for the children to exert all their energy so later the adults could rest and relax with a glass of wine by the pool while the kids swam.

Some years, we'd spend time with Mom in Orlando at her luxurious vacation property, surrounded by palm trees, beautiful pools, and bathtubs large enough to swim in. It was a place where she'd spend a week spoiling all of us. Mom bought it with the intent of one day passing it down to Sam and me to share, so we could continue our tradition of vacationing together, having fun, and making memories. We were two military families who didn't have much time to experience such lavish accommodations and were grateful for her generosity.

2009 - Sam and Renee during family vacation to Orlando.
Courtesy of Renee Nickell.

One thing was for sure, we always had a great time. Our children acted more like siblings than cousins. Oftentimes though, I'd catch Leah and Ben, both about three or four years old, holding hands. They were inseparable when we were together. I thought they'd grow up to be best friends, like I am with my cousins. Kylee was the oldest, usually responsible for at least keeping an eye on them, but antagonizing them as much as they did each other.

In 2010, we decided to go visit Sam and Robin in Virginia, tour Williamsburg and Busch Gardens, and then drive to North Carolina for Dad and Carol's 25th Anniversary. I was six months pregnant with our third child, little Ethan…and it was HOT. Walking around

Colonial Williamsburg in late May was a good way to be transported to that time period.

It was time to head back to the house. Mom and Donnie were visiting as well. They had planned this trip for quite some time over Memorial Day weekend. At the last minute, we got a call that Dad and Carol's anniversary party would be held during their stay with us. Sam had to relay the news.

"Mom…soooo, Kelsey is having a 25th anniversary party for Dad. We are all sort of expected to be there. We have to drive to North Carolina in a couple days. I can't really get out of it." Sam said half grinning as he knew this was quite an inconvenience. Mom and Donnie had driven all the way from South Florida, and now had to leave earlier than they had planned.

I can't say Mom was happy about this. "Well, Sam, we sort of had our plans first. I don't understand why this was sprung on us at the last minute," she replied.

"I don't know Mom, but that's not all…we have to make some food for the party too," he softly stated.

"So, you are saying during our last day with you, we have to spend all day in the kitchen cooking food for your father and his wife's party?" Mom said very plainly.

Sam was embarrassed and there was an awkward silence that filled the room. "Well…make a list of what we need. I guess we'll all just pitch in and help. The faster we can get it all done, the more time we'll have together for other things. No point in being upset over it," Mom replied.

Now, my late grandmother, my father's mother, had an infamous macaroni salad that the family either loved or hated. My mom hated it but knew how to make it just like Nan's. When Dad and Mom were married years and years ago, she'd make it for my dad. Sam and I grew up on that macaroni salad and we both loved it…and so did Carol.

"Mom, do you think you can make the macaroni salad?" Sam asked.

"Sure, Donnie can make the potato salad and I'll help with the macaroni salad."

Sam, Robin, Mom, Donnie, and I were all crammed in the kitchen, each with a task of helping to prepare food.

Fast forward, the party was over and Sam leaned into Dad, "Hey, guess what? Mom made the macaroni salad," he said with a smirk.

Dad smiled big, almost laughed and said, "Reeeallly?" A few chuckles escaped. I think we all felt the irony.

Sam, Robin, Gentry, Dad, and I were standing in Dad's kitchen long after the party guests had retired for the evening. Carol walked into the kitchen, opened the fridge and pulled out the massive bowl of leftover macaroni salad. There were A LOT of leftovers. Carol fixed a bowl and began to eat it while we were all making conversation. My dad, known for his ability to make people laugh, or to make light of any situation, said with excitement, "So Carol, you're not going to *believe* who made the macaroni salad!"

Oh my gosh. Did my dad just ask this question? We literally all just stood there with frozen expressions on our face.

Carol, the fork half way to her mouth after she'd already eaten some, nervously asked, "Who?"

Dad laughing says, "KATHI!" You could have heard a pin drop. We didn't know whether to laugh or leave the room.

Carol forced a smile and said, "Ooohhhh, really?" as her countenance shifted. The awkwardness increased, but we could all see my dad was so tickled by it. Carol left the kitchen and went back to her room.

Sam, broke the tension in the room and said, "Well, we better eat all the macaroni salad tonight because it will be in the trash tomorrow!" We all burst out laughing.

It became one of our favorite stories to tell. Sam had a quick wit about him and knew what to say at just the right moment to evoke laughter. It's not that he even tried to get people to laugh. He was just *that* funny.

Not All Fun and Games

Military life is wonderful. It's wonderful and hard. Depending on your spouse's career field, it can be great cause for concern. The worry, the dread, the anticipation of another deployment or TDY (temporary duty). Military life comes with its fair share of changes and it certainly is not for the faint of heart.

I remember when Gentry and I were stationed at Luke Air Force Base in Arizona. We were living in base housing and I befriended another spouse. Gentry was a late joiner, so by the time we got to Phoenix, he was 27 years old. The other airmen he worked with were in their early twenties. Not only was this gal a new spouse, married only a few months, but her husband was fresh out of basic training, too. At that point, I had already moved across the country by myself with a two-year-old.

I'll never forget what she said, "Renee, Mike may have to leave on TDY for two weeks to Nevada." I saw the panic cross her face. I stood there and waited for what came next. I could see tears well up in her eyes, "I just don't know what I will do without him for two weeks. I don't know how I will make it." I was dumbfounded and slightly annoyed. I tried to comfort her as best I could. *This girl is not going to do well as a military spouse.* New military spouses went through what every spouse goes through...the first TDY, the first deployment, the first move to a state or new country. You just get through it one day at a time, sometimes one hour at a time. You find out how strong and capable you are without your husband or wife.

Sam and Robin were no different from any other military family. Sam was an aviator and he was gone A LOT. Robin, a stay at home mom, wasn't interested in getting caught up with the officers' wives club, but always found a few trusted friends to rely on when needed.

When your husband is gone frequently, a marriage is often tested and tried. It just is, and if you don't have a solid foundation to fall back on, you won't make it. Gentry and I know that firsthand. We'd been married for five years at the time, and military life was harder on us than I expected. We'd been together since I was 19 and he, 21. We started our marriage with a baby and then moved across the country with no family support nearby to live on a military base. It was hard. We began to grow apart. I'm not sure where things went wrong, but Gentry served me with divorce papers after he'd returned from a TDY. I was devastated. I was a stay at home mom, without a college education. Now I needed to raise a five-year-old daughter on my own. I had no family, no money. I had nothing.

After I wallowed in self-pity for a month or two, I got my act together and decided to move off base. I started college, got my own rental home, and began my life as a single mom. Meanwhile, I prayed. I prayed and prayed and prayed and I will never forget the day when my husband decided he didn't want his child growing up in a broken family like we both had. He turned his life over to Christ and has been a changed man since. We tore up the divorce papers and started over. We've now been married over two decades.

Not every family is so lucky. Difficult times challenge every family. You can get in a place of denial which further perpetuates marital discord. What makes it even harder for military families is the repetitive reintegration from deployments. Gentry had been deployed to Afghanistan when he got the call that his seven-year-old daughter was critical. She had been diagnosed with Type 1 Diabetes and we had no idea until she got sick. They flew him home on emergency leave and then flew him right back into war again.

2007 - Leah walking to her daddy (Gentry) for the first time at his homecoming from Afghanistan. This picture made the local and base paper as well as the news. *Courtesy of Renee Nickell.*

His reintegration when he came home was HARD. We had to adjust to being a family again. Mom had been boss for four months. When Dad tried to step in, the children didn't want to hear anything from him. Add in a devastating new illness and the stress was unsurmountable. There is no guarantee that any TDY or deployment will go smoothly and most of the time, they don't. It takes a toll on marriages and every member of the family. Military families have to navigate through all sorts of hardships.

Reintegration back into a family is as important as preparation for deployment, yet somehow, it's very much overlooked. It is an oversight that may be the difference between life and death. Twenty-two veterans commit suicide every day. There have been soldiers that have killed their spouses, and then themselves, upon return from deployments. I've spent time with these surviving, hurting families and it's a devastating experience. There are organizations out there like the Renewal Coalition that help with the reintegration process which I'll touch on later.

Military life *is* hard. The strongest people I know are military spouses and the most admired military marriages are those that don't just hide from their problems, but they work through them. I know every circumstance is not the same. There are many times it is better for a spouse to leave the relationship for the safety of herself and her children. There are times when every effort has been made to save a marriage. It takes just as much strength to leave, but the families who decide to stay together through the hard parts and the good parts of military life are rarities and ones that should be praised.

Gentry is retirement age now from the Air Force and it saddens me that I've seen so many marriages end through the years. Somehow, Gentry and I made it. We stood firm on our vows and our faith in God. We relied heavily on prayer and the advice of our parents. Sam and Robin made it too. They made it through the hard times, which always made the good times so much sweeter.

CHAPTER 5

ASSIGNMENT AFGHANISTAN

Why go? I wasn't the only one asking the question. For two years, Sam had been training extensively with the 4th ANGLICO in West Palm Beach. He spent a lot of weekends, sometimes weeks at a time in Florida and visited Mom when he could. Many times, he brought the boys, Neil and Ben, with him. Mom thoroughly enjoyed her visits with Sam and the boys. Robin worked with the worship ministry at their church in Virginia Beach, Virginia. When Sam brought the boys to Florida on weekends, Robin's time was freed up to work on music with the worship pastor, Don.

Caesar, one of Sam's friends, ran into him at Virginia Beach. Caesar had been stationed with Sam in Lemoore, California, and Beaufort, South Carolina. Now, they were both in the Norfolk area. Sam had told Caesar, a fellow aviator, that he was deploying. Caesar, quite shocked, asked Sam about his reason for signing up for Afghanistan. Caesar had done an air and ground tour and felt like that was enough. The places he'd seen were about as close to hell as you could get without actually going there.

"Sam, what's up with this deployment? You know how dangerous what we do is!" Caesar stated boldly. Sam buttered up his answer and Caesar discerned there was something else to it. He felt maybe Sam caught the "grunt bug" otherwise known as combat addiction. The grunt bug is when marines turn into adrenaline junkies. They become addicted to the adrenaline of combat. It's like a drug. You get so close to death that it makes you feel alive. Sam had a good job and he was doing good work. Caesar didn't understand why Sam had signed up for a third deployment if it wasn't the grunt bug. Perhaps there was some other underlying reason Sam decided to go back to war.

Sometimes, Marines would rather be fighting in a war with their brothers than be at home. War has its own beauty. Combat simplifies your life and gives you very clear lines. Sometimes, your clear lines are two things: your rifle and your Bible. There are men and women in combat who are always clinging to both and if you don't have your Bible, you know who does.

It was the spring of 2011 and box after box had been shipped to Mom's house for Sam. She stored the large boxes in the back bedroom of her home. Sam had flown down for duty and he began to open all the boxes, filling the room with military equipment. It was all over the bed and the floor. Sam had waited a while before telling Mom he was deploying.

He was in the back bedroom of Mom's house with the door open. Mom had been out running errands and came in the house to see what Sam was doing. He stood at the doorway and she approached him. She looked past him with a stunned look in her eye. She asked, "Sam, what is all this stuff?"

Sam replied, "Um…well, it's for my deployment."

"What deployment?"

"Oh, Mom. Didn't I tell you? I'm deploying," Sam replied.

Mom stood there with her arms crossed, "Sam, where are you deploying?"

Sam paused for a moment and said, "Afghanistan".

Mom's heart sank. She didn't say a word. After the shock of hearing he was deploying, she said, "Sam, I would have remembered if you told me you were deploying to Afghanistan." Mom started to worry but didn't want to get freaked out about it. She pushed it to back of her mind. She did not want to dwell on it for the next six months before he left.

Soon after, Sam called me. "Renee, I have to tell you something," he said.

"What's up?" I replied.

"Well, I'm deploying to Afghanistan in the fall," he said.

"Oh? How long this time?" I replied.

"About eight months."

I remember thinking what a long time that was. I realized that there were military service members who would deploy for a year to fifteen months at a time, but I didn't want to think about Sam being gone that long. I wasn't worried about him, though. I knew it was part of his job and a possibility. I knew he'd be ok.

"So, you can't tell Robin, but I've been saving all my hotel points from my TDY's. I'm going to take Robin to Hawaii in June. We never really had a honeymoon and I want to surprise her. Just don't tell her," Sam said.

"I won't tell her, but that is so freaking awesome!" I was excited for him and I knew they needed this getaway together. Marriage is hard in the military and this would give them much needed time together to reconnect.

It was a week before Kylee's twelfth birthday and I could feel the impending deployment coming. I always love taking spontaneous trips. I mentioned to Gentry that I'd like for our family to go down to Mom's for Kylee's birthday. I thought we could spend time with Sam before he deployed and then take the girls to Disney World. I didn't so much as ask, but told Gentry that was what we'd be doing. I had a strong urgency to spend that time with Sam and I didn't know why. There were no bad feelings. I just knew we were supposed to go down there.

We spent the week together and Sam was particularly more quiet than usual. He would just glance over at me with his side stare and he'd grin a little bit, not really wanting to talk much. I knew he must have been carrying the weight of the world on his shoulders. I hadn't any idea how dangerous his job was, and maybe he didn't want to tell me. He was headed for some more training and so we decided to leave for Orlando.

He opened the door to leave and stepped down on the porch. I stood there, but I did not want to say goodbye. He put his arm around me and gave me a big brotherly hug. I was ignorant about his career field and would have hugged him a little longer had I known. It is still difficult to this day to think back on that moment and all the things I would have said. Perhaps I could have begged him not to go, as if that would have made a difference.

“Sam,” I looked him in the eyes. “Don’t go doing anything stupid over there. You just keep your head down and come home to your family,” I said.

He looked at me and smirked but didn’t say a word. No words of encouragement or comfort. No reassurance. The irony of my statement resonates with me today.

“I love you, Sis,” he said.

“I love you too, Bro.” I stood there and smiled as he walked towards the porch screen door. I did, indeed, know that he loved me.

2011 - Sam in Morocco preparing for his deployment to Afghanistan.
Courtesy of Jason Hartzell.

Sam called me one afternoon like he'd always done before. "Check your email," he told me. I opened my email and it was a picture his son had drawn. "Can you believe Neil? He made this picture for me and wrote a letter." I sat there, baffled at the maturity of his seven-year-old son.

Neil wrote, "I don't want you to go to war. If you die in the war, I will be angry for a really long time." Neil then drew a picture of little soldiers with a line between the two sides: evil and good. How could a child process war? A place so scary for a child to ever imagine and then to know that your parent will be going there, fighting against "the bad guys."

I brushed it off as a child's fear. I thought the letter was sweet and perhaps Neil's fears just needed to be comforted. I knew Sam would be safe. He was a Marine and he was deploying with the best of the best. What did I have to worry about?

Sam and Robin's anniversary was in November. That year, it fell the day after the annual Marine Birthday Ball. Robin left the children with her parents and she flew down to attend the ball and celebrate their anniversary together before he would leave just a few days later for his deployment.

Every military spouse's worst fear was to get the dreaded knock on the door for the service member in combat, but it is better to be prepared for the worst and hope for the best. Sam was very responsible and always planned ahead. Anyone who knew his love of Dave Ramsey knew this about him. Strategy is everything. He and Robin sat alone one evening and went over funeral plans if anything should happen to him. It's not a conversation any married couple wants to have, but it is a necessary evil when facing combat. Sam had even taken time that year, to adjust the driver's seat of his prized, beloved Mustang so Robin would be able to drive it while he was gone.

It was in that moment that Sam truly realized what he was leaving. This deployment was different and he had a bad feeling. I don't remember Sam crying very many times in his life, but he cried with his wife, not wanting to go, but knowing his duty to his country and his Marines. He would have never dishonored them. His goal at this point was to bring them all home safely to their families. There was no changing his mind. These brave Marines needed Sam's guidance and he knew that.

Sam and Robin reflected on the past, the mistakes, the failures, and the good times that all married couples face. They embraced their marriage and their commitment to each other, making plans for the future if, God willing, there was one. It was in that moment that Robin realized how much she wanted him to stay but encouraged him to go.

They attended the Marine Corps Ball and Robin was as stunning as she always had been; a classic beauty, mesmerizing in her blue ball gown. The next morning, they headed back to Mom's to spend their anniversary and their last day together there. Donnie made prime rib and lobster, his specialty, and a true gift to anyone who has the pleasure of sharing this meal with them.

(left to right) Sam, Jason Hartzell, Martin Castillo, Tom Cupo, Anthony Cuesta, Errol Miranda at the Marine Corps Ball.
Courtesy of Errol Miranda.

"Robin," Mom asked, "Why don't you stay longer? Just one more night."

"I wish I could, Mom. I want to, so badly. I'd have to call my parents since they are taking care of the boys and the dog." Robin walked back into the room where Mom and Sam sat. "I can't. I have to go." She sat on the couch, crying, sobbing really. She wished for just one more day with Sam before he deployed, but it wasn't possible. Her mother insisted she come home, which left us all to wonder what the urgency was.

The sun and all its glory came up over the horizon that Monday morning as Sam drove her to the airport. They held each other a little tighter than any other time in the past. Being an aviator, there had always been TDY's and training missions, but this felt…off.

Mom was headed to Sunrise, Florida, into work to finish up some things before she left for the day to spend it with Sam. She began to think about how grateful she was for all the time she shared with him before he deployed, but that day at work, all she could do was cry in her office. Unable to speak to anyone, she closed her door. Her co-workers, aware that her son was deploying, were understanding and left her alone. She remembered back to a time she gave him a verse when he was deployed to Kuwait. She later came across his Bible at his home in which he'd highlighted Psalm 91. That memory gave her comfort and she knew, ultimately, God was in control.

The rest of the day was spent sitting on their back porch watching movies together. Mom and Donnie had completely remodeled their home adding on a custom screen porch with a beautiful stainless kitchen and TV. This was a favorite place to be at their house, and it still is. Many conversations have happened over a glass of wine or a Jack and Coke, watching movies, and sharing stories. Sam even had his Marines join them for dinner on several occasions.

This porch shared the memories of first birthdays, Thanksgivings, Christmases, and many social gatherings with friends and family. When I think about home, I think about sitting across from Sam at that porch table as he leaned back in the chair, usually playing Angry Birds on his phone as he engaged in conversation, always with a Diet Coke placed in front of him.

This night, Mom sat there with Sam, her only son. Unsettled and fearful of the following day and the months ahead, their eyes met as she remembered him as a baby. There was no longer a baby boy before her, but a man…a Marine. Yet, all she could see was the baby she carried for nine months. She smiled as she remembered how at six weeks old, he started sleeping through the night and how she rushed to his room to make sure he was still breathing. At eighteen months, he was supposed to be napping (but hated naps). Instead, when Mom walked into his room, she stood there in horror as she saw the mess. He had painted himself, the walls, and everything around him in poop and tossed the rest on the floor. Those days were long gone, yet she wished she could go back, for just a time to slow it down. The years had gone too fast and now she was sending him off to one of the most horrific wars.

It's Time

It was Monday night and the evening was winding down. Everything was a blur. The day went by way too fast. The sun had said goodnight and the moon shone brightly in the clear night sky. Mom walked into Sam's room and said goodnight. He still had some items to load up in the car before he left the next morning.

Mom didn't sleep well that night, fearful that she would somehow miss her goodbye, as if he wouldn't have wanted to wake her. As she tossed and turned, she'd look at the clock each time, dreading morning. 4:30 came quickly and she arose to make Sam some coffee in the kitchen. With a true mother's heart, every time we were home, Mom always made us a cup of coffee each morning. Sam was a simple coffee man while I always got an expresso with steamed milk.

Sam had already been up loading the car with the rest of his personal belongings. They stood in the kitchen for a few last moments. "Mom, it's time. I have to go," Sam said hesitantly, not wanting to walk out those doors. This was a place of comfort and safety, a place he was always at ease. Regardless of those little marital spats and quarrels between Mom and Donnie, the ones that became funny over the years, we just learned to roll our eyes and smile because we knew they couldn't live without each other despite the times they had gotten on each other's nerves. This place…it was home. The military had taken both our families many places over the years, but Mom's home was always a place of comfort, healing, and love. It was a refuge for us.

Sam reached toward her and wrapped every bit of his arms around her small frame. He completely embraced her. After a few moments, she began to pull away, but he wouldn't let her. She leaned into him once more as they held each other on that quiet, early morning. As she held on to him, she felt it may be the last time.

They finally let go. Mom let go first. She knew Sam had to go. "Sam, you have to go or you'll be late."

Sam looked into her eyes, "Mom, I love you."

"I love you Sam," she replied as a tear rolled down her cheek.

She walked him out onto the porch in her pajamas and robe, and watched him prepare to leave. He stood at his car door, turned around, and looked back at his mother and smiled. He got in his car and pulled away as she stood there watching him disappear into the distance. He rounded the curve down the road and she burst into tears, crying uncontrollably, wanting him to come back or to force him to stay. It wasn't possible. He was gone.

That week, Mom had a hair appointment. Her friend Lydia has been doing her hair for years and they not only shared a long-standing friendship, but also their faith. Lydia owns a quaint little shop around the corner from Mom. The shop was set in a rowhome with beautiful professionally manicured landscaping. It was like a scene out of the Truman Show and each small business had a place for the owner to live upstairs.

Mom walked into Lydia's shop. There was no greeting today. Mom looked at Lydia and said, "I have a really bad feeling about this deployment," as she rushed over to sit in her chair. Lydia tried to comfort her as best she could. Mom just sat there almost in a daze, trying to reconcile with these unsettling feelings she had, yet could not explain.

She sat there in her chair, again thought of Sam and what a wonderful baby he had been. She chuckled under her breath and thought about how he had tried to kill me on occasion. Sam apparently didn't like me very much in the beginning.

She remembered one day when they were at my grandparents' home in Pennsylvania. Mom was changing my diaper as my dad stood next to her. Sam quietly climbed in between them and bit my toes and ran away. They didn't even see it happen. I let out a blood curdling cry. Mom thought she had stuck me with a diaper pin. She completely stripped me looking for any sign of injury. She looked down at my foot and saw Sam's teeth marks. She knew in that moment she could never leave me alone with Sam again, at least until I grew on him a bit, and, thankfully, I did.

It's amazing how the simplest daily activities can become the fondest memories. My mom, reminisced, as she felt in her heart that she would never see her son again. She pushed down the fear and hoped that was all it was.

Mom sat in that chair at Lydia's shop and thought of how thankful she was to have a son like Sam. This little boy who once dreamed of becoming an F-18 fighter pilot, overcame obstacles, and never gave up when times were hard. Sam kept pressing forward through the hard times to reach his dreams. He had a beautiful family, a wife Mom loved more like a daughter who gave her two precious grandsons whom she adored. Ben, the youngest, was a spitting image of Sam.

Sam had always given so much. It was just his nature to give himself again to his Marines, be there for them, guide them, and bring them home. That was his final goal. She felt peace knowing she had a part in raising this honorable and noble man, her son. He knew what he had to do and he was an expert at it. Surely that spoke for something. That was enough to keep him safe...that and her faith. Why did she worry so?

The Holidays are Here

The holidays were just around the corner and we were preparing for a trip to North Carolina to spend the week with Dad and Carol. This year was Dad's 60th birthday. It was certainly something to celebrate. We were all a little heart sick that Sam would not be there for the party. With the holidays approaching, I thought about Sam often, over "there", with God knows what to eat for their Thanksgiving and Christmas meals.

Thanksgiving Day came with more food than the eye could take in. Later that evening, Dad checked his voicemail. There had been a missed call. It was Sam. No one had heard the phone ring. He left a message wishing us all a Happy Thanksgiving and promised he would talk to us all later. Down in South Florida, Mom had missed Sam's Thanksgiving call as well. They were at a dinner with friends at the Jupiter Golf Club. She looked down and saw a number pop up and didn't recognize it. A moment later, she realized it could have been Sam. There was a voicemail. She listened. It was her son's voice. She immediately began to cry. She had no way to reach him.

Robin was the only one who answered the phone that day. Thankfully, Sam had reached her.

The party planning had begun for dad's birthday bash because there would be plenty of friends and family gathering to celebrate. Laughter was in full swing. My half-sister, Kelsey, my half-brother, Daniel, and I were goofing around as always. We'd seen a video online and had planned to replicate the skit, and send it to Sam for Christmas. Unfortunately, time was short and we kept putting it off. We decided that'd we'd do it the next time we got together, perhaps in the spring.

Carol delegated some of the party planning to everyone. She was always great at making sure the smallest details were special. She loved pictures and always tried to make sure she gathered as many pictures as she could from the family. There was a pile of pictures all over the kitchen table. Carol had asked me if I'd help hang them all around the house. There were pictures of my father's life from the time he was an infant until grown. Memory after memory lay in front of me. I sat there organizing all the pictures, carefully attaching each one and taking in memories of the past.

The week was wonderful. My father, always a source of laughter, made that time extra special and Carol always threw the best parties. Even though Sam was deployed, we were all looking forward to the approaching Christmas. Just a month prior, I had purchased a model car for Sam for Christmas. It was a '67 Fastback in a shade of green close to his '68. Right before he left for his deployment, I told him I'd hang on to it and give it to him the next time I saw him. And so, I was counting down the days.

CHAPTER 6

OPERATION ENDURING FREEDOM

"Man Down"

(Based on the testimonials of Detachment India)

Sam's position as a Forward Air Controller would mean he would view a live feed. Anything his team needed logistically or by means of aircraft support, they would route it through Sam and he would make sure they had everything they needed. Sam was the Detachment Officer in Charge, so he spent his time behind a screen on a large base calling in air support. He wouldn't be exposed to the constant threat of IED's (Improvised Explosive Device) and direct contact with Taliban fighters that his men were, or so I thought.

"Sam became part of the Fire Power Control Team and supported the British and Royal Marines as the Fire Power Control Team Leader and the Joint Terminal Attack Controller (JTAC) but was still the

Detachment Officer in Charge (DOC). He went out with the Royal Marines and acted as that liaison, the one who would provide any kind of military aircraft, artillery, or anything else that needed coordination to shoot the Taliban. Anytime the Royal Marines would see action in the field, Sam was out there with them. He would take the radio and get aircraft support on station and persecute the enemy with the aircraft. He was a lot deadlier with the radio than anyone else was with a rifle," Staff Sergeant Ralph Perez explained to me during a conversation we had.

2011 - (left to right) Jason Hartzell, Sam, Errol Miranda, Jim Winters during deployment to Afghanistan.
Courtesy of Jim Winters.

SSgt Perez further volunteered that Sam was such an expert at what he did, he didn't need a monitor, but he likely had one. Sam had a small screen attached to a radio and he saw what the aircraft saw. Hypothetically, if the unit got into a firefight while on patrol, Sam would connect to the aircraft by radio. At that point, it was just Sam and the aircraft communicating. Sam would tell the aircraft which direction they were taking fire, how many bad guys there seemed to be, and where their location was. He would talk the aircraft onto the enemy. The aircraft acknowledged that they saw the same thing Sam saw, the enemy, and they knew where shots were being fired from. Once it was established that there were no (or minimal) civilian casualties and it was safe to drop bombs, they went ahead and did so with Sam's approval.

Sam's team was only allowed to take two guys from their four-man team on patrol at any one time. Sgt Jason Hartzell had been going for three to four days straight and was getting hit with firefights every day. December 13th, 2011, Sam recognized that Jason was under a lot of fire and getting tired.

Jason shared this story with me. "I met Sam over three years ago when we began training for our first deployment together. I was fortunate enough to be on his team ever since then. Looking back on our time together, the impact one person can have on your life is rather remarkable. I'm not speaking strictly in regard to his impact on my military career. He did teach me everything I know about being a JTAC but his lasting effect on my life was more than that. He was truly his brother's keeper, always concerned for my well-being and willing to help me out in any way possible.

"I believe he viewed me more as a little brother who he needed to keep an eye on, but also knew when to let me make my own mistakes. His sense of humor was witty and random. Although, his best jokes were the ones he thought were really funny, but weren't. He'd have that ridiculous grin on his face and you couldn't help but laugh anyways. I always loved his tales from college because I was also a Sig Tau and it made my college stories sound slightly less bizarre. He was one of the most enthusiastic storytellers I ever met, rarely could he deliver the punchline before he started laughing. Anyone who knew Sam well, also knew the nod. I'm still not quite sure what it meant.

"In the day leading up to our mission, we were driving to a different base to stage for the next day. Along the way, we came across a catastrophic accident where a man and child on a motorbike had been struck by a car. The father was horribly injured and I began treatment. Sam, ever eager to help, began assisting me applying splints, a C-collar, and eventually a needle decompression in an attempt to re-inflate the man's collapsed lung. Once I had stabilized the patient, I looked over and saw Sam tending to the child who had a broken leg. He had a calming presence about him by nature. He truly cared about people in a way that is rare these days. He was the epitome of a Christian man; imperfect but strived to be the best person he could possibly be.

"On the first day of the mission I went out with our British counterparts and Afghan task force we support on foot. The rest of the team, including Sam, was with our truck near the edge of the town to provide support and submit air requests among other things. While patrolling, I quickly realized the town was practically empty. As we approached a compound, machinegun fire broke out and I threw myself into the nearest ditch I could find and began to return fire. I could hear the snap of the rounds over my head and the dirt kick up from impacts all around me. Needless to say, the situation was pretty intense and not ideal. The gunfire died out, we accomplished our tasks and concluded the first day out.

"I met my team at the truck, prepared our sleeping bags, and talked about what happened that day. Sam told me we were switching for the next day and Winters would go with him as his radio operator. He'd be the JTAC on the ground and I'd do the air requests and surveillance. After a pretty brutal day, I knew he was going to do that. I knew he couldn't stand another day of hearing me get shot at over the radio and watching it via a live video feed from the aircraft overhead. It just wasn't in his nature. On the second day, the volume of incoming fire was higher compared to the previous day," said Sgt Hartzell.

One of the reasons I never worried about my brother, Sam, was because I knew, as a commander he would never be in the direct line of fire. As the Forward Air Controller, he was making the calls, on a base…safe. This day Sam relieved Jason of his position to give him a break from the previous day's fire fight.

Sergeant Jim Winters recalls, "I remember working with Major Griffith the day before on the video scout as he was trying to track the bad guys that were constantly harassing the British patrol Jason was supporting. It was very frustrating to get the PID (Positive IDentification of the target) necessary to prosecute the targets. I had a strong feeling Sam would want to go out the next day. After debriefing the day's events with Sgt Hartzell, Sam informed the team he was going to go out. He asked me if I was up for it. Of course, my answer was, 'Yes, I am always up to go with you.' Sam and I had already done a couple of patrols together on previous missions and I always felt privileged to be working with the boss.

"I had not known Sam as long as some members of Detachment India, but I had come to trust and admire him during our time, more than any other Marine Corps officer I have ever served with. He was not just my OIC (Officer In Charge) and Team Leader but had become a good friend. Not many on the detachment had a wife and kids like we did. We would talk about our family whenever we got a break on patrol. My youngest is right between Sam's two boys.

"While on patrol that day, we came up on a large compound and moved in to establish a position prior to moving across to our next objective. I will always recall our last conversation while we were taking a short break. We were eating from our British rations that had these packets of smooshed up fruit and I commented on how it was like baby food. Of course, Sam said he had never eaten baby food, and I was in disbelief. He proceeded to talk about how Robin and he made all the baby food for their boys. He was a man who definitely held to his beliefs and truly loved his boys.

"When it was time to move out, the Afghanistan force led the patrol out of the compound, followed by some Brits and then Sam and me. A portion of the Afghan and British Forces remained in the compound to provide over watch. Right outside of the compound was an irrigation ditch that made it a little difficult to get going. Once the patrol was across the ditch, we paused for a moment before heading out.

"We were maybe 200 feet outside of the compound and all hell rained down. The gunfire was as accurate as it was intense. I dropped behind a small mound and returned fire," Winters explained.

Sam was covered in Kevlar protective gear, virtually everywhere except his face. As the gun fire got more intense, Sam knew that he needed a visual on where the fire was coming from. Instinctively, his plan exposed himself to the enemy fire, which would give direction of where they needed to return fire.

Winters went on, "Sam was standing next to me when the firing started. I turned around to check on him and he was down. I moved over to him, calling out, 'MAN DOWN, MAN DOWN!' I could see that he had taken a single round to the face and was unconscious. Still under fire from the Taliban position, the British corpsman quickly arrived and assessed his wound. There was no pulse or respiration. After wrapping the wound, three other British soldiers and I evacuated Sam back to the compound and prepared him for the medevac."

Sgt Errol Miranda began, "I was in the MRAP with (Jason) Hartzell. We were checking aircraft and battle tracking. When the radio erupted with traffic that, 'A Yank was hit!' It took me and Hartzell a second to register it. I will never forget looking at him and him looking at me."

He continued, "Then all the training that Sam put us through kicked in. Harztell started requesting air assets for the Brits and I started putting together the CASEVAC and reporting what was going on."

"It took the Brits a few minutes which actually felt like an eternity to let us know it was Sam," Sgt Miranda said solemnly. Under the midday sun, on a cold December day, the 14th to be exact, their leader had been killed instantly by a Taliban sniper, by small arms fire in the Helmand Province, Afghanistan. After being medevac'd to Bastian Hospital on Camp Leatherneck, Sam was officially declared deceased.

"At the end of the day when we finally egressed from the village, I had to ride with the British forces back to a Forward Operating Base to rendezvous with Errol and Jason. It was the longest, darkest ride of my life. Alone, my team leader, my boss, my friend, gone," Sgt Winters expressed.

"The day Sam was killed is the most helpless I felt since my mother passed. From that point on, we regrouped at the vehicles and had a long silent drive back to Camp Leatherneck. We arrived in time to see him shipped home. Winters, Hartzell, all the Brits, ATF (Afghan Territory Force) and I rendered hand salutes and tears. We lost our boss, mentor and good friend," Sgt Miranda stated.

Detachment India deployed with twenty Marines. Sadly, they returned with only nineteen.

"Sam's team, Winters, Hartzell, and I felt the best way to honor him was to take the fight harder to the enemy and use the year of hard quality training Sam put us through as our guide on how to perform in the battlefield. Without his caring to ensure our success in battle, we may have lost more. His legacy with ANGLICO will be that of a true mentor and friend." SSgt Miranda stated.

Sgt Hartzell offered, "His decision to pull me off the frontline and put himself in my place ultimately led to the tragic loss of his life. I would have been in the same position, doing the same things he was if he had not swapped with me. His choice to assume the risk and face the danger in my place subsequently resulted in saving my life that day. To his family I would like to say on behalf of mine, we are eternally indebted to you for your sacrifice. If it wasn't for Sam, my family would currently be going through what you are now. It is a debt we will never be able to repay. Your family is in our prayers. To Sam, our team is back in the fight. We're Marines and that's what we do. After all, we were trained by the best. I will miss you, I love you, and I will never forget you, Sam. Semper Fidelis," Sgt Jason Hartzell 4th ANGLICO, USMC DET I, JTAC.

Sam's heroic actions that day ultimately saved a lot of lives. He was posthumously awarded the Bronze Star with Valor. Our mother, father, and Sam's widow received his Purple Heart.

Where was the Knock?

Mom was driving home from work on December 13th, a Tuesday. Traffic on I-95 was the usual parking lot. Inconsiderate drivers congested the freeway. Mom, enjoying the quiet, suddenly heard a voice within her. "He's not coming home." She immediately began to pray. Being a Christian, she was very much aware of spiritual forces, and felt like perhaps this was a fear tactic. She asked God to keep Sam safe and protect him. She asked Him to give her peace and take away those thoughts of fear. She didn't think much of it the rest of the day.

The following morning, December 14th, she left for work at 7:30 a.m. with an hour-long drive ahead of her. She was running a bit late. About fifteen minutes from work, the phone rang. She noticed it was a local area code. *Whoever it is must wait until I get to work.* Mom had just unlocked her office door and didn't even get to put her purse down before the phone rang again. Perplexed, she wondered why she'd received two calls within such a short time at this early in the morning. She looked down and saw it was Robin's phone number.

"Mom? Are you sitting down?" Robin's voice was trembling.

"Robin, what's wrong?" Mom said urgently.

"Two marines just came to my house."

Her brain started spinning and she knew what was coming but was not prepared to grasp it.

"Sam is dead," Robin said, crying.

"Are you sure? Are you sure? Are they sure it's not a mistake?" Mom was yelling at her.

"Mom, it's no mistake," Robin said through tears.

"Oh God, I'll call you back," Mom dropped the phone. All she could do was scream. Everyone could hear her.

A young woman came running to her office. "Are you ok, Kathi?" she asked.

"My son is dead! My son is dead!" Mom screamed as she was hanging onto the side of the wall. Hunched over, shaking and crying uncontrollably. The young girl stood there in shock, not knowing what to do.

Mom closed the door. She frantically realized she needed to call people. She needed to get home. She was in no shape to drive and she lived an hour away. She listened to her voicemail and heard Colonel Mallard, "Ma'am you need to call me right away," he said. What would he have said to her on the phone while she was driving?

Mom's neighbors looked out the front window and saw two Marines walking up to the house. They knew immediately what was happening and got on their knees and started praying for her.

She called Colonel Mallard and he said, "Do you know why I'm calling you?"

"Yes, my son is dead," she said.

"Where are you?" he replied.

"I'm in Sunrise."

"Do you want us to come there?"

"No, I'll get a ride home with my boss," she told him, eager to get off the phone.

Mom called her manager. She was trying to say something, but Katelyn couldn't make it out. "I can't understand you, Kathi," Katelyn said.

"Katelyn, my son is dead," Mom told her as best she could, almost hyperventilating.

"Are you in your office?" she asked.

"Yes, I'm here." Mom replied.

"Just stay there, we'll get you home." Katelyn said. The Vice President of her department drove her home. Dad called. She picked up the phone.

"Kathi, do you know?" he said, voice trembling.

"Our baby is dead. Do you know the details?" she asked.

"Yes, I do, but they will tell you," Dad replied.

Mom got home and all she could do was sit on the sofa and cry. Her best friend, Mary, was trying to console her. "Just remember, God is with you."

"It doesn't feel that way!" Mom cried.

She got up and walked down the hallway into the back bedroom where Sam's dress blues hung in the closet…his dress blues that he had worn to the Marine Corps Ball just a month before. Tears streamed down her face. She unzipped the bag and rubbed her hand down his jacket and took it off the hanger. She clutched it close to her and held it all day.

Hours passed and people came and went. Mom's pastor was there and then she began to get briefed by the Marines on what happened and what was about to happen.

Mom realized she had to pack, but how could she? How does a mother pack her suitcase to receive the remains of her dead son coming home from war?

Mary sat with her as Mom tried to rest. Night fell and as she was about to leave, she noticed the local Channel 12 news truck drive into the driveway.

"I can't talk to them," Mom said.

"You have to do this for Sam," Mary told her.

The news stations kept coming. Just as one would leave, another would pull in. Mom couldn't handle it anymore. She needed to be alone. She took Sam's jacket, laid on the floor that night, and held it as she cried until she couldn't cry anymore. Sleep would have been a welcomed escape from the nightmare. Nothing could have prepared her for what was next.

A new day came, and the sun rose again. Unfortunately, she didn't awaken from this horrific dream. *Did that happen?* she thought to herself as she began to awaken from a moment's rest. As the reality set in again, she realized she had to fly to Dover this morning.

CHAPTER 7

THE IMMEDIATE AFTERMATH

It was like any other day in December with the anticipation of Christmas arriving in a few days. Christmas has always been a special time for our family, and it's among my fondest of memories as a child. It has always been my favorite holiday and the most special. Not only because it celebrates the birth of our Savior and regardless of the quandaries that happened as a child, Christmas was the time when we were all happy. We were all a family.

The kids had a couple more days of school before Christmas break started. The weather was perfect and I was eager to get started with my Christmas shopping. There is a little white clock we've owned since 1997 which has been in many bathrooms in the different homes we resided in during our military life. It is still in our bathroom today. My daughter, Kylee, was twelve at the time and was helping her five-year-old sister get dressed that morning. I had just showered and gotten myself dressed. My hair was fabulous and

I was looking in the mirror to put on my mascara with long even strokes. I can so vividly go back to that time. Almost as if I can feel the mascara brush in my hand and my face up close to the mirror, so not to smudge my make-up. I couldn't wait to head to Destin for the day. The sun was shining and it was a beautiful 68 degrees.

The phone rang. The clock sitting on my night table read 7:34 A.M. Gentry was on his way to Duke Field Air Base for work. I looked at the phone and saw it was mom's number. *Why on earth would Mom be calling me at 7:30 in the morning?* Smiling, I picked up the phone. I was always happy to hear from her. My perky, happy, not so normal morning voice said, "Hello Mom."

All I could hear was hysterical crying. "Renee!" She spoke my name in the most haunting manner. She was mumbling something I couldn't make out.

"Mom, I can't understand you!" I said, deeply concerned by her tone.

"Renee, your brother…"

"Mom, I can't understand what you're saying!"

I heard, "Your brother," and thought to myself, *Oh God, he's lost his legs, but at least he's alive and coming home. Ok, we can deal with this.* In that split-second, all I could fathom or imagine was he had been injured. I just needed to understand what had happened and how bad it was.

"Renee…"

"Mom, tell me what happened!" I began to panic as she was trying to talk.

Her voice was quivering so badly, she could hardly get it out, "Youuur broothheerrrr! He's dead."

"MOM! What do you mean?"

"Your brother was killed in Afghanistan last night."

I fell to the floor. It felt like my soul had just been ripped out of my body and all I could do was scream. I dropped the phone and hunched over holding my stomach as if my organs were coming out of me. It felt like I had been standing on a mountaintop and the ground below me collapsed. I picked up the phone and told my mother I had to go. I didn't know what to do but scream and cry.

Kylee came running. "Mom, what happened?" She knew. Years later she told me she knew the moment I started screaming.

"Mom, tell me what happened," Kylee pleaded.

"Uncle Sammy. He's dead. I don't know what to do! I don't know what I'm supposed to do!" I continued screaming and crying, barely able to get my words out.

Kylee immediately took Leah and baby Ethan into a bedroom. Leah asked, "Why is mommy crying?"

"Shhhhh, It's ok, Leah. You both just stay right here. Everything is ok. I'm going to go call Dad but I will be right back. Don't come out. Just sit here in this corner and I'll be back." She said in the sweetest, calmest little voice.

"Ok Sissy," Leah said.

Kylee came out and grabbed my phone to call her father. I gripped the countertop, almost knocking over the whole stack of Christmas cards that were stamped and ready to place in the mailbox that morning. Those were the last Christmas cards I've ever sent since that day. "Daddy, Uncle Sammy died."

Gentry on the other line, still on his way to work, turned the car around in the middle of the highway, "I'm on my way home, Baby Girl."

Kylee started calling every close friend we had. Our neighbor could hear my screams from their house, even with the doors shut. She ran over and took the baby with her. She later told me she still is haunted by the screaming she heard that day. Her husband, also in the military, was deployed at the time. There are no words for agony. There are no words of comfort for what I felt. Friends began showing up, one after the other, bringing food and embraces. Dallas packed for our children. Andrea and Bonnie brought food. We had to go. *What do we do?*

Mom called me back, both of us in complete shock. "Renee, they are flying us to Dover. We have a casualty officer here who will accompany us."

"Mom, what do I do? How do we get there?"

My mom answered, "I don't know. You may have to drive." I could tell she was as disoriented as I was and couldn't make clear decisions at that point. She was just trusting her Casualty Officer to guide her through the process.

The phone rang again and it was my stepmother, Carol. "Renee honey, I can't believe this has happened," she cried. "They are flying us to Dover but we don't have all the details yet and I believe your sister and brother will be able to fly too, but I'm not certain. We'll have to bring Aunt Mabel with us because there is no one here to care for her."

I replied, "Ok, just be safe, and we'll leave as quickly as we can."

"Just drive safe honey, and kiss those kids for me," she said.

"I will. I'll see you soon. Love you, goodbye."

Friends were coming and going from my home that dreadful day, helping wherever they were needed, mostly offering love, support, and condolences. My best friend, Joy, rushed over just as fast as she could. A dear friend, Neressa, brought her pastor since at the time we were transitioning between churches.

Pastor White came to me, a kind, gentle man with white hair, and a big welcoming smile. I imagined him to be the sweet grandpa any kid would be lucky to call theirs. His gentle disposition and warm embrace held me as he prayed for me. His eyes filled with tears as he thanked me for the courageous act of heroism by my brother. Reaching into his pocket he pulled out four $100-dollar bills. "Take this." I could see it was everything in his pocket. I didn't know what to say. "Thank you" didn't seem like enough. I just hugged him and cried.

Gentry, the kids, and I started our journey to Dover, Delaware, a place I had never paid much attention to until then. 7000 of our military had been killed since 2001 and I had never paid much attention to where they came home. And so, we drove. Parents, spouses, and children were able to fly to Dover. Those are usually the ones listed on the casualty report as members to notify. If you are not on this report, there is no assistance for you, unless you are a dependent, adolescent sibling. Typically, the Casualty Officer will assume the parents will notify the siblings. We drove from the panhandle of Florida to Dover, DE with $400 to our names given to us by a perfect stranger who turned out to be an angel.

2011 - (left to right) Andrea, Renee, Dallas, and Bonnie, celebrating with Christmas lights. Taken four days before Sam was killed.
These women came to my rescue that fateful day.
Courtesy of Renee Nickell.

Dover

On a single military income with three children, we didn't exactly have a stacked savings account to make a sudden trip to Dover, Delaware all the way from Florida. But the Lord provided us a way to get gas, lodging and food until we could get to the Fisher House. The Fisher House is a foundation that serves the military community and its families. They provide beautiful homes to house family and veterans during times of medical treatment or the death of a military service member. They cover all expenses and meals during a stay, regardless of length.

Oh no! The Fisher House. My mom called me to remind me to call and reserve a room, as she was still en route from the airport. Just hours a way, I called and a young man answered the phone.

"My name is Renee. I am the sister of Major Griffith. We'd like a room for our family please."

The gentleman replied, "Ah, Ma'am. I don't think we have any more rooms. How many are in your party?"

I answered, "Two adults and three children. One is an infant."

"Ok, Ma'am, I'm going to have to check on this and see if you can stay here. Ma'am, we were told not to expect any other family members," he said.

"What? What do you mean? We've been traveling for two days to get there and we weren't even able to fly in."

"I'm sorry, Ma'am. We weren't expecting another sibling. When we were trying to make accommodations for everyone coming to the Fisher House, we specifically asked if Major Griffith had any more siblings so we could plan accordingly. The woman we gave the last block of rooms to told us Major Griffith did not have any other siblings. We were not prepared for your family. I am incredibly sorry for the misunderstanding. Let me check what we have."

I was furious and hurt. Who would say such a thing? I couldn't fathom why, at a time like this, someone would have said such a thing? Perhaps, it was a misunderstanding, but I doubted it.

"Mrs. Nickell, we have one room available with two twin beds."

"Well, if that is what you have, that is what we will take."

"I'm truly sorry Mrs. Nickell. If we'd known you were coming, we would have reserved more rooms and had others double up."

"I understand. I am just so grateful to be able to stay there with my family during this time."

"Yes, Ma'am. Your room will be ready to go. And Ma'am, I'm terribly sorry for your loss. Please let us know if we can do anything for you to make your stay as stress-free as possible, under these tragic circumstances. When you get here, please make yourself at home."

"Thank you, sir," I said solemnly.

After two days without any sleep and sixteen hours of driving, we finally arrived in Dover and were shown to our room.

Families are complicated. I've heard it before, many times, that families fall apart after a military death. I needed to let go of the fact that my two half-siblings each had bedrooms with queen beds and our family of five was stuffed in a room with two twin beds.

I do not think it would have been too much to ask for our family to be accommodated during an already stressful time, but Aunt Mabel would have never heard of it. She wasn't exactly what I would call a gracious woman and she ruled the roost when she was around. I loved my family and I was grateful we were all together. I'm certain if they had known what was said, they would have made better arrangements. It wasn't until later that week that someone confirmed to me that they overheard Aunt Mabel mention to the Fisher House employee that Sam had no other siblings coming (while we were en route to Dover).

We were all pacing around the Fisher House waiting on his remains to be flown in from Germany. This was a beautiful home with several private rooms. It held enough for multiple family members to sleep comfortably, each with their own bathroom. The kitchen was big enough to feed an army. Beautiful white cabinets and a sleek marble countertop island nearly eight feet long decorated the elaborate kitchen. This was a kitchen that would be found only in one of the finest homes. A huge playroom with endless toys filled the space beside the kitchen and gave a sense of comfort to children who just lost a parent.

"Renee, can you come here?" Carol said to me. She led me into their room. She pulled out a gift box. "I want you to have this. We were able to stop on the way to the airport."

I opened the box and found a beautiful locket. It was a stainless heart with a diamond in the center. I was speechless. I opened the locket and inside were two pictures of my brother, Sam. I knew Carol had picked it out. She was always so thoughtful when it came to gifts. "Carol, I don't know what to say. It's lovely." Tears began to swell in my eyes and I hugged her and my dad. I put the locket on and I still wear it today.

Service members and top officials came and went. They informed us of every detail and what to expect next. Several never left our side. My father and I stood in that kitchen, eyes locked, still in complete shock from what had happened. *What became of our lives?* I was also beginning to wonder what would become of our family. We questioned each other without having to say a word. His look was so solemn as if he didn't know what to say. Such sadness stared back at me from his eyes, and I knew things would never be the same. He began to say something. I didn't know if I should comfort him or he comfort me. He said, "I just don't know how your sister is going to get through this. She really looked up to him."

What? What the hell did he just say? I was furious. I couldn't take one more second of the way he babied her anymore, or how he always completely disregarded my feelings. I nodded, tears in my eyes. I had to get out of there. I should have known to expect it.

I fumbled for my key to my room. I quickly let myself in and went straight to the bathroom. Rage filled my heart. I locked the bathroom door, thankful it was completely private. I stared at myself in the mirror. *Who was that person I stared at?* Tears began to flow down my cheeks as I yelled at myself in the mirror, "HOW WILL *SHE* GET THROUGH THIS? HOW WILL *SHE* GET THROUGH THIS? HOW CAN YOU ALWAYS BE SO INSENSITIVE TO ME? I'M YOUR DAUGHTER TOO! SCREW YOU! YOU ALWAYS FORGET ME! HE WAS MY BROTHER FOR 34 YEARS! HOW THE HELL WILL I GET THROUGH THIS? HE WAS THERE FOR ME WHEN NO ONE ELSE WAS!"

Tears now dripped off my chin faster than rain drops off the eaves of a house during a torrential downpour. I pounded my fists on the counter until I thought I would break them. I couldn't take the pain, the heartache. My brother's dead body was on its way home, and I couldn't even get an acknowledgement from my own father that I, too, needed him.

I gained my composure so I could return to the main living space with Gentry and the kids. I had to pull myself together. I was used to this. I coped. I got hurt, I'd cry, scream, and then I pushed it down into the depths within me and moved on. *Get yourself together, Renee.*

I went out and saw my dad and mom sitting in the corner of the living room. I walked over and pulled up a chair. The three of us sat there and something made us laugh, talking about Sam. A surreal disbelief came over me that he was coming home in a flag-draped coffin. I tried to visualize the scene in my head to prepare my heart. It grew quiet after the laughter dwindled as if we were a bit ashamed to have had that moment of joy.

I looked at my dad and lost my composure. "It was Sammy." I began to cry, "THIS IS SAMMY! THIS IS SAMMY! HOW CAN THIS BE SAMMY?" I looked at my dad and mom's faces across from me.

I'd never seen pain so great on my parents faces. Growing up, it was always Sammy and Renee as if we were one word. Now Sammy was gone. They returned my emotion, each with tears streaming down their faces. My father sat there, one arm crossed against his body, and the other hand covered his mouth, all of us unable to face the grim life shattering reality of what was about to take place.

I never questioned whether Sam would be injured and I certainly never believed he would be killed. To me, he was invincible. Had I known how dangerous his mission was, I would have hugged him a little tighter during our last goodbye before he left on deployment.

The casualty officer who was holding our hands through this entire process gathered us together. He began to tell us what was about to happen. "Sam is almost home. I need each of you to hear me. What is about to happen will be the worst day of your life. This moment, when he comes home will be worse than finding out he was killed. I need you to understand this. This will be the hardest moment you ever experience. This is when it becomes real."

I stood there dumbfounded. I couldn't absorb what he was saying. How could this day be worse? I looked around at the stone-cold faces that surrounded me. Everyone tried to comprehend what he had said.

"It's going to be very cold. Make sure you're bundled up. It's almost time."

The "Angel-Flight" Home

The bitter cold reminds me of the hole that now resides permanently in my heart. We boarded the transport bus in what seemed to be the dead of night. Perhaps it was. There was nothing but pitch darkness except for a red flashing light inside. There was no sound, only sniffles from us all trying to hold back our tears in the bitterly chilled air. We tried not to make eye contact. Gentry stood beside me and gripped my hand, letting me know he was there for me through it all.

The moment we would face with anxious anticipation was about to become reality. It must have been a three-minute bus ride that seemed like an hour. The vehicle came to a stop as it was parked on the flight line. From a distance, we saw the silhouette of the C-17, the back open and just the tip of a flag draped coffin. A roped area and chairs were lined up as a reminder to the family to stay in our designated area.

The family unloaded from the bus and walked toward our assigned area. The cold didn't even faze us on that December night, only one week before Christmas. An ambulance with red and white lights approached the C-17. The ambulance pulled up behind the plane to prepare for the transfer of his remains. *His remains.* I hate that term, as if there is nothing left of him, but his remains. I realize it's a blanket term used for all those killed in action. Some come home whole and some come home in pieces. It is a cold, hard, reality. Nonetheless, our loved one was gone, taken from this world, never to grace our presence again.

The back door of the C-17 was open. I could barely see the back tip of his flag-draped coffin. I was unable to believe that Sam was in that box. The transfer began. Seven young military men exited the aircraft. Every tear that was withheld, now started to flow and the display of agony commenced from each of us. I became unaware of anyone around me. All I could do was envision him lying there. *Was he even in there?* Yes, it had crossed my mind many times. *What if it wasn't HIM*? My mind began to use coping strategies against my own will to decrease the pain of the reality.

These airmen were well trained. Every movement was in complete synchronicity. They must have practiced time and again to give families the last gift they are able to receive: a dignified transfer of their heroes. That moment was the last moment we shared the same space. That moment was it. We would never again be in Sam's presence again until heaven.

"He is not fit for viewing," we were told. Sam wished, when he passed, to be cremated. I'm still not sure how I feel about not seeing him one last time. For years, part of me mourned that I did not have that tangible reality of seeing him one last time, solidifying the fact that he was, in fact, gone. The other part of me was grateful that I remember him alive and vibrant, full of life and laughter.

I wish I could remove the image from my mind of what I *felt* he looked like. No one really talks about the post-traumatic stress a family endures in these moments, trying to cope with such great a loss. It was done. The transfer was complete. Now it was time to begin the process of moving through the funeral, navigating Christmas, and learning to survive. Not live, survive. I had entered survival mode. We all had.

2011 - Sam's remains return home on American soil, Dover AFB.
Courtesy of Renee Nickell.

2011 - Sam's dignified transfer of remains, Dover AFB.
Courtesy of Renee Nickell.

2011 - Sam's remains being loaded onto the ambulance.
The hardest part was watching it drive away.
Courtesy of Renee Nickell.

The Memorial

It was time to gather and celebrate the life of my brother, a life well-lived. Family and friends came from all over the country. People kept ushering in, one after the other. We had previously planned on having Sam's memorial service at his home church, but it had to be moved because so many people were expected. Upwards of 1,000 people filled the church to remember Sam.

The family stood in line to enter the church. Sam's best friend, Matt, Robin, and I sat up the night before going through the details. The next morning, Gentry and I, and our kids were patiently waiting outside the doors to go into the sanctuary. There was no set order really, but ahead of us were Robin and the boys, my dad and mom and stepparents...and Aunt Mabel.

Aunt Mabel turned around and Gentry happened to be in her wrath path that morning. She leaned into him, pinched the little bit of fat behind his arm like she'd done a million times to me and Sam as a child. She gritted her teeth and stared him dead in the eyes, "You WILL walk behind me into the sanctuary."

Gentry quickly pulled away, and without hesitation replied, "Mabel, I wouldn't think of walking into that church before you."

Aunt Mabel then leaned my way and whispered in her sweetest tone as she always had done, "Hey, sweetie, um, we're going to try to get you on the front row with the rest of the family, otherwise, you will have to sit on the second row." I knew exactly what she meant. She never fooled me.

"Oh, it's ok, there is room. I'll be sitting on the front row."

"Yes, well, we will do our best, but just know you may have to sit in the second row. I just don't think there is enough room for your family," she said and smiled, obviously concerned there wouldn't be room for us, much like at Dover.

She certainly had no idea I had been present for the planning of the seating, and she need not be so concerned about such things. Yet we both soon learned that I didn't really need a seat on the front row. I had actually been placed on stage for the majority of the ceremony, much to Mabel's chagrin.

On our way to Dover, I had written words about Sam. I never realized they would be spoken to a crowd of 1000 people or spoken at all for that matter. When I entered the church, I opened the program and saw I would, in fact, be giving the eulogy. I felt panic inside. I would have to bare my heart in front of all these people just days after my brother had been killed. By the grace of God, He gave me abundant peace at the moment I needed it most.

I entered the sanctuary with my family and I took the stage with a few of Sam's closest friends and Marines. I was in a complete cloud. My brain was in a fog and I sat there praying that I wouldn't cry as I spoke. I didn't want to fall apart in front of all these people. I blocked out every sound and sat there in disbelief that I was about to speak at my brother's memorial.

I remember I sat there listening to others speak about Sam. Then my mom, my dad...both in a state of shock, took the stage holding back tears. I was in utter disbelief as I watched them speak at my brother's memorial service. It still seems surreal to this day. It's almost as if I need to shake the memory out of my head as I feel my pulse increase thinking back to that day.

2011 - Renee giving Sam's eulogy at his memorial service, Virginia Beach, VA; surreal seeing this photo. It doesn't seem possible.
Courtesy of Renee Nickell.

I approached the podium and looked out on the people who had gathered. I couldn't believe how many people were there to honor my brother. There was an entire section of impeccably dressed Marines. Cousins I hadn't seen for years had come. Friends who shared his childhood and his college days looked up at me. There were aunts, uncles, grandparents. *How can this be happening? Is this a dream?* I began to speak:

> "I've been thinking about what my brother would want me to say at a time like this. Knowing him the way I do, in his humility, he wouldn't want me standing up here telling you what a great person he was. We all know what a great person he was. From a young age, my brother was a person of faith. I don't know at what age he accepted Jesus, but I do know that even before he knew Him, Sam still had faith. Maybe at the age of five, he didn't know what it was…but he did know for sure that he had enough faith and determination to set a goal and reach it. My brother gave 100 percent in everything he did. From taking the fine grains of green colored sand to make grass along one of the largest train set displays I've ever seen in our basement to becoming an F18 fighter pilot twenty years later.
>
> "There are so many descriptions of my brother, I don't have enough time to repeat them all. I could stand here and tell you about all his accomplishments; however, that doesn't make a person who they are. Sam was so much more than his accomplishments. While I'm very proud of him and his successes, those things won't pass with you into the next world. So, I will tell you about what does…character. I will tell you that Sam was honorable in every aspect of his life. He was the most honest, trustworthy person I've ever known.
>
> "I've been pondering so many characteristics about him over the last few days. And this is for sure; He was a man of God.

He cherished his family, his wife, his sons, and his parents. He could make us laugh until we cried. What I wouldn't give to hear that laugh again. He adored his mother. There is no greater loss to a mother than losing the child she brought into this world. He was an outstanding son.

"Mom, you did an amazing job as a mother. You taught him forgiveness, compassion and love. You taught us how unforgiveness leads to bitterness. I always admired your willingness to quickly forgive.

"My dad instilled in him honesty, integrity, responsibility, determination, perseverance, and a hard work ethic. Let's not forget both your infectious laugh and endless supply of jokes. Dad, Sam was a picture of your outpouring into him. He was just like you. I know you are so proud.

"That being said, I'd like to share with you the characteristics of Sammy, as I call him. He was honest, he was brave, and he was courageous. He was honorable, he was hard working and the most positive person I knew. He was encouraging, and he never spoke a harsh word about anyone…ever. He was loving, levelheaded, and rational. He was generous. He was the one to always seek out for good solid advice. He was responsible and trustworthy. He was funny yet knew when to be serious. He was an exemplary Christian man. He WAS the picture of a true Christian because his role model in life was Jesus. He was never judgmental or critical.

"He never lifted himself above another or glorified himself. I never heard him boast about his life as an officer and a pilot. He had every right to, but he saw himself the same as everyone else. He was what Paul talks about in Philippians 2:2-4. 'Let nothing be done through selfish ambition or conceit, but in lowliness of mind let each esteem others better than himself. Let each of you look out not only for his own interests, but also for the interests of others.'

"My brother always looked out for the best interest of others. He even died doing so. Sometimes we think that to do something great for Christ, we need to be a world-renowned evangelist, we need the title of pastor, or we need to donate millions to orphans in Africa. Christ humbled himself even unto death. He washed the feet of his disciples during his last supper before his crucifixion. He continually showed us throughout scripture how to live our lives. My brother's life was a witness. It was a witness to how a true man of God should live. The honor my brother received in his life was because he was humbled before the throne of God. God honored him with favor so he could touch a world.

"Isn't that why you're here? Because Sam touched your life in a special way? You see, there is one thing that is certain in this life: we will all die. And if Sam wanted me to say one thing, it would be this, 'Jesus is the way, the truth and the life and no one comes to the Father except through Him.' He would say, 'If you want to see him again, if you want to share with him in Glory, then you need to have the faith to believe that Jesus saved you through the cross and all you have to do is accept Him, as Sam did.' That's it. Sam didn't fear death, because he knew where his destination was.

"1 Corinthians 15 tells us that 'Death is swallowed up in victory. O death, where is your sting, O grave where is your victory?...Thanks be to God, who gives us the victory through our Lord Jesus Christ.' I've gained a new perspective on this life. I feel closer to heaven than ever before. I no longer fear death. 1 John 2 states, 'To not love the world or the things of the world. If anyone loves the world, the love of the Father is not in him. For all that is in the world, the lust of the flesh, the lust of the eyes, and the pride of life is not of the Father but is of the world. And the world is passing away, and the lust of it; but he who does the will of God abides forever.'

"Sam was not consumed with the lusts of this world. He lived a humble life. My brother has made me re-evaluate my priorities. What impact do I want to make in this one life I've been given? You see, if his death doesn't change me, if it doesn't make me want to be a better person, a better Christian, if it doesn't make me want to pursue God with everything within me, then I will never have peace with his passing. I will always ask why.

"I stand here today, ready and willing to recommit myself to the cause for Christ. The Father says, 'There is no greater love than to lay one's life down for his friends' (John 15:13). Sam was everybody's friend. God sent his one and only son, the King of Kings, to live as a pauper, a servant, with no title... to die for you and me. My brother left his family, his wife, and his sons. He willingly left his homeland, fought with everything he had, and laid his life down for you...and for me. What greater love is there? What other greater display of Christ is there than to live humbly as he did and die to save another?

"He loved being a Marine. He loved his friends and I'm so sorry for your loss as well. Sam was a devoted husband and father. I pray you realize that a piece of you didn't die, but lives. A piece of you lives in heaven. A piece of Sam lives on in the lives of Neil and Ben. Sam's memory lives on in all of us.

"Sam, I look forward to seeing you again. I know I will. I have the blessed assurance that I will see you again by the simple faith of accepting Jesus. Let there be nothing wicked found in me when The Good Lord calls me home. Let us all strive to be more like Sam."

Sam's 34-year-old widow, Robin, took the stage with hands shaking. Her beautiful brunette hair was softly placed into a bun and she gently swept her bangs out of her face as she approached the podium. It's not typical for the widow to speak at funerals, but she had a gift for Sam she wanted to share with the world. She grabbed a microphone. The music pastor, Don, whom she had a close working relationship with at their church and who was a great source of spiritual comfort for her when Sam was deployed, graciously agreed to sing with Robin the most beautiful rendition of "I'll Be Home for Christmas," Sam's favorite Christmas song.

Sam did make it home for Christmas, just not the way any of us had ever planned. I still think of that day whenever I hear that song.

Each shot of Sam's 21-gun salute shook me to the very core. BOOM! BOOM! BOOM! Even now, when I hear those sounds, I'm transported through time. My mind goes blank and I begin to stare into the distance, frozen in time. I'm reminded of Robin receiving that folded flag to forever remind her of the great sacrifice her husband made. It was and is a reminder of the honor, loyalty, courage, and devotion Sam had to her, his family, his Marines, and his country.

2011 - 21-gun salute over Sam's prized 1968 Mustang Fastback
at his memorial service, Virginia Beach, VA.
Courtesy of Renee Nickell.

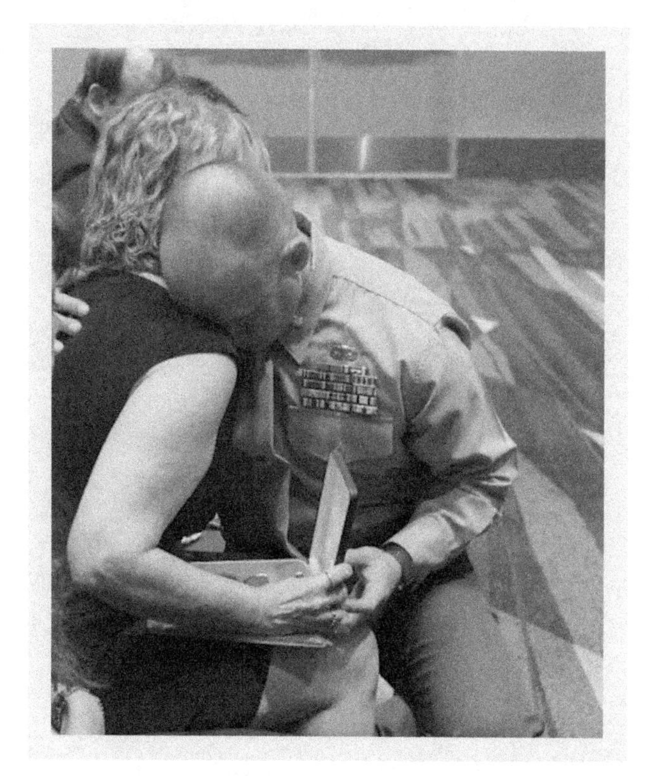

2012 - Mom receiving Sam's purple heart during his
memorial service in West Palm Beach, FL.
Courtesy of Renee Nickell.

CHAPTER 8
YEAR ONE

Christmas

After the funeral, we sat around opening all the beautiful cards sent to Robin. We each had a stack. As I looked down, I saw the handwriting and realized two of the letters were from Sam. Sam had mailed each of his sons a letter for Christmas. "Robin, there are two letters here from Sam." I said, my voice trembling.

"Well, open them!" she said excitedly.

It was surreal to sit there and open cards from Sam after his funeral. His words came off the paper as if he were still alive. Christmas would never be the same. It's the little things we cherish, like a handwritten note inside a card, homemade muffins made by caring neighbors, a song that brings flashes of memories to mind, new and old friends that will always be there like family. I still wonder how we will go on.

My phone rang. "Renee, we were trying to get into your house to assemble the kids' Christmas gifts and the alarm went off," my friend Cyndi stated. "I'm so sorry to bother you."

Before Sam had been killed, I had ordered one gift for each child off a website and had them shipped to the house. Leah got a dollhouse, Kylee got a keyboard, and little Ethan got an activity center. We needed someone to call who would put together the only (three) gifts we had bought for our kids for Christmas. So, I called one of the most dependable people I know, Cyndi and Garry Johnston, and they came to my rescue.

"The cops are here," she said.

"What? The cops?"

She continued, "Yes, we are trying to explain to them that we are, in fact, assembling a doll house and not stealing it."

"Ok, let me talk to them." I kindly explained our dilemma to the nice officer and got the alarm turned off. Cyndi and Garry spent the next eight hours assembling a dollhouse so our children would have presents for Christmas when we returned home. At that point, we didn't know when that would be.

As Christmas Eve Day dawned, we continued to walk around in a daze. There seemed to be a lot of confusion developing within the family, things being said or done which we attributed to the continued shock and grief. Nothing made sense. I went along with anything Robin needed or wished of me. On that day, what she needed was sushi which was part of their Christmas Eve tradition. We all piled in the van and ate sushi in an empty restaurant.

I sat across from Robin and stared at her. I knew she was the absolute love of Sam's life. He would have done anything for her. I was so thankful she shared this tradition with us. I was thankful for our friendship and wanted to do whatever I could to be there for her and the boys. I wondered how she would make it as a young widow and I questioned why God would allow this to happen to our family.

I began to hope that at any moment, a duck would be placed in front of me with its head chopped off as it had been in the movie *A Christmas Story.* I think we all would have welcomed the comic relief.

"You know, we don't have any presents for the kids," I said to Gentry and Robin.

She replied, "I didn't have time to shop for them. How could I?"

"Well, I know they will all understand. We'll make it as special as possible." I stated.

The next morning, the doorbell rang and we saw a Marine in his dress blues at the door. "Good Morning, Ma'am," the Marine stated to Robin. "We are deeply sorry for the loss of your husband and we'd like to provide Christmas for all the children." At that moment, two Marines began unloading gift after gift into the house. There were more gifts than I had ever seen. We all began to cry and the children were bursting with tears and excitement. Our entire family had been contributors for Toys for Tots for many years. I had never imagined we'd ever be a recipient. We thanked God for a Christmas miracle during the hardest moment of our lives.

2011 - Our Toys for Tots Christmas after Sam's death, Virginia Beach, VA.
Courtesy of Renee Nickell.

We made it through Christmas and it was time for Sam's ashes to make their way home. Brent and Matt drove up to the house in a blue Mustang. It couldn't be more fitting for Sam to return to his home than in a Mustang. We gathered outside to watch as they pulled up, sharply dressed in their dress blues. They held back tears as they approached us with a brass, engraved urn in hand. There was no emotion shown. They held their bearing, as any Marine would, and kept their tears from flowing down their faces.

They handed the urn to Robin and she cradled the remains of her dead husband in her arms. Little Ben, her son was overheard asking his older brother, Neil, how their daddy fit in such a tiny little box. They began to discuss that the government must have "Willy Wonka'd" him, shrunk him down so he could fit in there. How can you explain to a six-year-old and a seven-year-old the process of cremation? We, as adults, so callously think nothing of this process. It was in that moment, we all knew these young men had a lot to learn about death and grief. I was heartbroken for them. How many other families must face this very thing?

As Christmas and New Years passed, the days ran into each other. It was time to head home. I feared the journey because I didn't want to leave Robin and the boys. It was easier being with them than having to face the reality of grieving alone.

Feeling Close to Him

I always loved riding with Sam in his car...or cars. He'd had a plethora over the years since he was sixteen. Sam called me up one day "just to chat" like he normally did. He said, "Renee, you're not going to believe this. I got a steal of a deal on a car. I bought a 1993 Honda Civic for $500."

Of course, I was thinking two things. One, this is like, the mid-2000's. Why on earth would a fighter pilot who was an officer buy a $500 car to drive? And two, did it even run?

This was probably during his Dave Ramsey stint when he was preaching to me about getting out of debt, tithing, spending less... all that big brotherly advice. I had gone to visit him when I was pregnant with my third. He had to make a run to the commissary for groceries for the evening's festivities and asked me to tag along. I had the pleasure of riding in his 1993, green, Cadillac of a car...Uh hem, Honda Civic. I gently made fun of his car, especially when he had a restored 1968 Mustang Fastback in the garage at home.

We always had a great time together when it was just us. He would usually give me his big brotherly advice and I'd try to impress him with whatever I was doing at the time in my life. "So, Sam, if you ever want to get rid of this car, I'll take it off your hands for you. Just let me know when you want to get rid of it," I joked. He'd begin to brag about its fuel efficiency and how he got around just fine in that car.

The irony of it all is that I never dreamed I would actually become the owner of the very car we laughed about on occasion. I took my last ride with Sam in that car, and now it's mine, graciously given to me by Robin. I see it every day, parked in our garage and think of Sam. Now, my husband, who is also an officer in the military, drives this $500 car to and from work every day. He ditched the $50,000 F150 and drives a 1993 Honda Civic to work that may or may not have running A/C at the time. Sam would be proud.

2011 - Sam's 1993 Honda Civic, still referred to as "Sam's car". We are hoping to one day fully restore this car in Marine Corps red as a tribute to Sam.
Courtesy of Renee Nickell.

There are times when I just go and sit in what is known as "Sam's car" and I feel so close him. I place my hands on the steering wheel, knowing that he at one time drove it. It was a long time before I would let anyone drive it. As a matter of fact, when we were driving it home from his funeral, I insisted I would drive it behind my husband who had the three children in the minivan. We had about a 700-mile drive after Christmas, 2011. About three hours into the trip it began to rain and Gentry stopped suddenly because traffic had come to an abrupt stop. I hit the not so great brakes, slid, bounced off the guardrail and hit my own minivan.

We ended up driving the Honda to North Carolina to make an impromptu stop at my father-in-law's. Once we got it to the shop, the mechanic assessed the damage. "Ma'am, this here is going to cost about $1000 to make it drivable." I couldn't even be mad. All I could do was laugh. It cost us $1000 to fix a vehicle that cost Sam $500. I *had* to laugh about the entire thing. The left front fender was dented as well as a nice dent along the door. I still smile when I see that dent in the door that I never repaired. There are certain things that must remain the same, so I can continue to remember. Then there are those things that must change, so I can learn to move forward.

How can you move forward when you don't even know what direction to turn? I look back and wonder, how do you take that first bite of food after a tragedy? How do you go grocery shopping or buy new shoes?

We made it back home safely and I remember standing over the kitchen sink reaching for the hand pump to squirt soap on my hands. I stared at the bottle of soap. I loved that particular fragrance of hand soap. I never bought drug store soap. I always bought the specialty soaps. It just made me feel good to have my most favorite scent at every sink. As I stared at that bottle of soap, I thought, *Why the hell* *did I take the time to buy this stupid bottle of $9.00 soap? Who cares what kind of soap you have? I'm never setting foot in that store again.* Nothing mattered. Something as little as a bottle of soap made me ponder the importance of material things when my brother had just been tragically killed in combat.

It wasn't just the hand soap, it was everything. How would I ever feel close to him again? There I was in the house where I found out my brother died. It seemed all the happy memories had disappeared. I stood there in a daze.

In my own experience, I looked for ways that Sam somehow was helping to guide me. I wasn't one of those that could "talk to the dead". I felt awkward even thinking about it. Some people do it with ease. I never could. I knew nothing I said to him would change the fact that he was dead and gone. I wanted to talk to him. I just couldn't. I wanted to pick up the phone and talk to him in real life even though I knew that wasn't possible.

It can be quite therapeutic for people to talk with their loves ones after they die, but all I felt was worse…and I didn't even want to try. To me, talking to Sam felt like a lie…not really knowing if he could hear me or not. That did change slowly as I began to heal. Now there are times when I'm driving down the road alone, I just talk to him as if he were sitting next to me. However, for me, it didn't happen overnight.

CHAPTER 9
THE HARD STUFF

I walked into my closet to sort through old clothes. I tried to find pretty much anything to occupy my brain. Clothes would get sorted by color and by season. Items I hadn't worn in a year got tossed. If I was waiting to lose weight to fit into something, I threw it out because by the time I lost weight, there would be new clothes in style and I would have a reason to shop. Not that I felt like doing that anyway.

As I sat on that closet floor, a small box with a folded flag caught my eye, as if it had just been thrown back there to be found at just the right moment. I picked it up and read the word on the outside "TAPS". *TAPS? What's that? Where did this box come from?* It turns out, after a military member is killed in action or dies during active duty service whether by suicide or health conditions, The Tragedy Assistance Program for Survivors (TAPS), sends a grief kit to all immediate family members. When I opened that box in my closet, a world opened to me that I never knew existed.

Just one month after Sam was killed, I met Bonnie Carroll, the founder of TAPS. I was invited to the 2012 Big Miracle movie premier in Washington D.C., along with actress Drew Barrymore. I was in a fog, pretty much out of my mind. I dragged Mom with me to this event. My mom wore a bright red hat.

We walked up to this woman at the after party. Bonnie immediately hugged us after we introduced ourselves. I may have been in a fog, but I remember her greeting, "Who are you here for?" She wanted to hear about my brother. I can tell you, most people don't want to hear about him. Grief is hard to talk about, but she broke that barrier immediately.

"I'm here for my brother Sam. He was killed in action one month ago," I said.

I remember her shocked look on her face. "Did you say, one month?" She looked at us and grabbed us close. I didn't understand. Isn't that what everyone is here for? I didn't understand her shock that we were there. Now I do.

Now that I am years beyond the shock of his death, I look back and wonder. *How did we do that? How did we get on a plane, fly to Washington D.C. and attend a movie premier just one month after Sam was killed?* I know how we did it. We absolutely were not in our right minds. Like many grieving families, I somewhat blindly made decisions. Some I regret and some I don't. That decision ended up being everything Mom and I needed in that moment.

I am grateful to TAPS for the friendships I have gained over the years. My children loved their uncle like a second dad and I am grateful that even though they are not technically Gold Star Children, they are able to participate in Good Grief Camp for Gold Star Kids to help process their grief as well. There were many times I couldn't help my children process their grief and TAPS stepped in where I could not. I am forever indebted to them for the compassion they have shown our family.

2012 - Renee and Mom at the TAPS Big Miracle movie premier, one month after Sam's death; A wonderment at how we did that.
Courtesy of Renee Nickell.

I had no idea the shock I felt could give me the ability to travel several hundred miles just one month after Sam had been killed and would continue for much longer than I had realized possible…or wanted. When you're in shock, you don't know you're in shock. You can't rationalize it. The shock and grief continued in ways I couldn't have imagined. I was 34 when my brother died. My children, 12, 5, and 1, needed me and I wasn't mentally or emotionally available for them. I was there physically, but I wasn't *there*. I slipped into a deep depression after about four months. I didn't know what was happening except that I did not know how I would go on. My heart was so shattered, I wanted to die. I didn't want to die because I wanted to escape my pain, or to be with Sam in heaven. I just felt like I couldn't go on. It's a feeling so indescribable that often happens when you experience such a traumatic loss.

I went to church one Sunday, feeling completely defeated. Four months is not a long time in the grand scheme of a traumatic death. Most think one to two weeks is acceptable and you should be going back to work and enjoying your daily activities. After all, the funeral gives "closure". UGH! I hate that word and I want to punch people who tell others they need closure. This is where we learn to give God's grace. I still couldn't even believe Sam was dead. There were times I wondered if it had been a mistake. I started fantasizing that perhaps he was still alive, only deployed and on some secret mission. How could I have closure if I couldn't even believe he died?

The church service was over and I spotted a friend. I thought, *Oh, God, she spotted me. I'm going to have to talk to someone this morning. I'm going to have to put on that fake damn smile.*

Krista said, "Hey honey, how are you? How are things going?"

I said, "I'm holding up. I'm doing alright."

She said, "How's your family?"

I replied, "Well, everyone is sort of falling apart right now. Relationships are becoming strained. It's just really hard."

She added, "Oh." Pause. "Well, is that what is keeping you from moving on?"

I didn't even know what to say. How do you reply to such an ignorant statement?

Well, I will tell you what I did. First, I forgave her for her ignorance. Second, I became thankful that she had no idea what it was to lose someone so close. Third, I went home and screamed into my pillow, even more frustrated that no one understood. Four months had passed and I should have moved on by now? I spent 34 years with my brother and I should have moved on after just four short months? I was with him longer than most marriages last these days and *I must move on?!?* Sam was part of me and when he died, a part of me died too. The depression only got worse the more I realized how alone I really was.

Love isn't a door that you can close and call it closure. Time isn't a Band-Aid that heals a broken heart. Grief isn't a 12-step program to be accomplished. The deep, deep love for a son, a brother, a father, or a husband isn't a feeling we just need to "get over and move on" when they die. Sam's life was never some whimsical story to get over, as if we could simply delete the painful scenes from his death. We only learn to function as best as we can with the big gaping hole until one glorious day, we are reunited and made whole again. By God's grace, we will finish the race stronger than when we started and be forever grateful for the time we had.

Our lives were falling apart. I couldn't be Mom anymore. I didn't know who I was. I was fighting with everyone around me. My typical MO was to push people away and that's what I did. Whether it was intentional or not, that was the result. Gentry and I were battling through horrible, horrible fights.

My mom and I were fighting. We went four months without speaking. I thought she never wanted to speak to me again after the horrible things we said to each other. I recently found a card she had written me for my birthday during the time we weren't speaking. I had wondered how she could ever forgive me and I was too prideful to reach out to her. This card, this beautiful card that was such an expression of a mother's love towards her child…was everything I needed in that moment.

I was crying when I reached her on the phone. I begged her to please forgive me. I didn't want to lose her but I didn't know how to navigate through this grief either. I knew her love for me was unconditional and regardless of how much I pushed her away, it was her gentle words that were so powerful and healing for me during that time. She didn't search for the right words to say, she just spoke what was in her heart. There is a difference. There is a difference between just saying something to say it and saying something meaningful from your heart. That made the difference for me.

Unfortunately, I didn't know how to do that myself. I didn't know how to be a vessel of healing in my other relationships. I was so broken. I didn't know how my words affected others, especially when I was placed in situations that required confrontation. I was not good at confronting anyone. I still have to work on it today, because I go into shut down mode if I'm not careful. My historical response is to shut down, confront abruptly, blame, disconnect, and lose the relationship. It has been this way my entire life. So, after Sam died, I couldn't even function in my healthy relationships, not to mention those that I had to let go of in order to heal emotionally.

A dry erase calendar hung on the wall in the kitchen. Behind it was the pantry and to the left was the garage door. No one could miss that calendar. That was the purpose. Everyone knew what mom had planned for the month on that calendar. I'd used it for years. The calendar was dated December 2011. The last date circled was the 14th. There was nothing on the calendar for the remaining of the month. Just that circled day, December 14th, 2011.

I had forbidden anyone to touch it. It hung there untouched for nine months. I wasn't ready to take it down. I couldn't erase that circle. How could I erase what happened? That was the day our world ended. Our happy family, our family vacations and beautiful Christmases... the future I imagined for our families...gone. That date stared at me in the face. Every. Single. Day. It was a reminder for the rest of the family too, but I didn't think they had to process Sam's death like I did.

One day, I walked into the kitchen and the dry erase board...was erased. I screamed, "WHO DID THIS? WHO ERASED MY BOARD?" I began to cry, "That was MY BOARD!" I wasn't ready. I wasn't ready YET!

My oldest, Kylee, the one who had to step in and be an adult on the day Sam died, screamed back. "I DID! I DID IT!" She started crying and screaming at me, "I'm so sick of looking at that calendar every single time I come in this kitchen! I'm so sick of being reminded of that day!"

I was so angry. I was angry she erased it and I was angry at myself for not even considering the children's feelings. How could I? I was blinded by my own pain. I left the house. I couldn't deal with her. Besides, it was MY board. I was the mom and it was MY brother. She had just erased the most significant day of my life without asking me first. Her anger grew and so did mine. We were mother and daughter, yet, we couldn't understand each other. I retreated to a childlike state, feeling like she was just being defiant or disobedient to me and I couldn't see the bigger picture.

She's always been stubborn with a fierce passion, but I felt that Kylee was over the top this time. I couldn't deal with her. I was done! Obviously, the counseling she was receiving wasn't helping. It was her, her, her. SHE needed to be more respectful of ME. I was her mother, after all.

The Weight of the World

I don't even know how we made it through that time. I was an emotionally absent mother and I felt like my child was just being spiteful and disobedient. Fighting ensued almost daily and I felt like I was a teenager again, living with my parents. I was raising my daughter but it felt like I was dealing with…me. It was as if I had been transported through time and I was raising myself as a teenager from a bird's eye view. I didn't understand it at the time. I also didn't know how to comfort her or help the situation. The tension was so thick.

Perhaps that's what happened in my own childhood. My parents took my rebellion as defiance and never could understand their part in my pain or accept it. It is not easy to self-reflect and I was still grieving. So, how could I? I was fighting with everyone. It felt better to get it out, fight, and just withdraw into hiding so I didn't have to deal with anything or anyone anymore. I couldn't deal.

Death was an uncomfortable subject, one that I did not want to discuss. I allowed others to suck me into their mess so I didn't have to deal with my own mess. But then, I couldn't get out of their mess either. That further complicated things with family and I continued to pick people off like I was shooting them dead. Boom, boom, boom!.They're gone. I didn't have to deal with any of them anymore. I wanted to be the hero and protect those I felt had been wronged, but I couldn't confront in a healthy way. One by one, I lost more than just Sam. Perceptions were skewed.

During that first year, my grieving became overly-sensitive and I could not make sound decisions. There is a good reason why counselors advise, "Do not make any major life decisions in the first year following the death of a close loved one." You forget conversations. You remember things incorrectly because you are not fully "there". Your brain is in a state of self-protection. This is a time when many families begin to fall apart. I tried to protect others; however, I had no business doing so. Everything began spiraling out of control. Eventually, I also became estranged from my father and half-siblings.

My half-siblings and I always had two different perspectives growing up. I couldn't understand theirs and they couldn't comprehend mine. It did not mean either of us were wrong, it just meant we had different perspectives, different upbringings. I don't fault them for that. Now, Sam was gone, the peacekeeper of the family was absent. At the time, I was relieved that my father and half-siblings were not in the picture. It was easier with them out of my life. I no longer had to care about how they were coping. I no longer had to deal with their grief or drama. I could just focus on my own issues.

As a sibling, I felt left behind. Ceremonies, gifts, remembrances… were for the parents and the spouse. I've spoken to people who have experienced the death of a sibling in Vietnam and they still ache in a way that no one can explain unless you've been through it. It stays with you always. Others say, "I understand," so flippantly. No…no you don't "understand". You don't know what it's like to lose your sibling and then watch on the sidelines as everyone honors your brother and invites your parents, while you feel like a third wheel, unworthy really and undeserving of participating in Gold Star events. And, if you actually do get invited, it is because you are "just a sibling". It's an overlooked grief.

Even crying on the phone with my mom made me feel so guilty. Somehow, I should have been stronger. What I ended up doing was stuffing all my pain. People never said the right thing. No one understood. It's as if I had a flashing red light above my head that said, "My brother died, so please avoid any meaningful conversation with me." I'm not saying this is what everyone thought, but it's what I FELT people thought because I had learned to keep my guard up at all times. I often didn't (and many times still don't) even want to bring up the subject of my brother because I didn't want to make other people uncomfortable. But how did this help anyone? How did that bring meaningful relationships into my life?

So, I slowly stopped talking about him unless it was a special Remembrance Day. I'd post something on social media and get a few sympathy "likes". I didn't want sympathy. Maybe I just wanted validation that it was okay to still grieve. I mean, I still went to pick up my phone at times to text him something silly. That's when the memories really hurt. The pain decreases over time, but missing him never went away. That never healed.

Flashbacks. No one warns you of PTSD. The flashbacks…hearing my mother tell me that my brother was killed. They were paralyzing. I would freeze in my tracks and stare off into space. If Gentry or the kids were nearby, they would ask, "Mom? Are you ok?"

I wonder what my face looked like when I got the flashbacks. They happened at the strangest times: while I was brushing my teeth, walking into church, or driving down the highway. The flashbacks brought a rush of anxiety for no reason at all. They are less frequent now and I can usually shake myself out of it after a few seconds, but I'd be lying if I said I don't ever have them anymore.

There are some things that I have a hard time putting into words. Things like pain, remorse, regret, grief, hurt, and brokenness. On the flip side, I can talk about joy, laughter, blessings, and thankfulness rather easily. I know, for myself, it is quite easy to hide pain behind a smile. I think I've become a pro at it. Something happens though, when God brings you through the really, really hard stuff. Eventually, you begin to smile because you really want to smile. This process was not overnight for me. It took time...a long, long time.

Quite honestly, I've come to realize that you can be a recovering alcoholic...and talk about it. You can be a recovering addict, or maybe you've been through a broken marriage, or abuse...and you can talk about it. But, start telling people that your brother died in war and they look at you like you have the plague...and then, most likely, they'll avoid you all together. You have this grief stigma attached to you. Whenever anyone talks to you, they are thinking that they don't want to talk about your grief, they don't want to talk about your dead brother, and so, they don't really talk to you much at all. Grief becomes the elephant in the room whether intentional or not.

I say this happens in the beginning stages...when the grief is constant. Now, I am at a place where I want to talk about him. I want to remember him. I want others to remember him. Perhaps others may see it as "still grieving" or being "stuck in my grief", but I'm where I should be. It's my journey and we each have our own path.

I do not have a degree in psychology, my knowledge comes from lots of experience dealing with complicated relationships. My experiences as a child caused me to consistently be on the defensive. I was blamed first and never asked questions later. Ignorance was bliss, I suppose. I dragged that into my future relationships. By the time I was an adult, I had the burning passion and lack of tolerance for any form of deceit or injustice. I became self-righteous versus god-righteous. I say this because I did not understand how to communicate to those that hurt me. My defense was to attack and retreat. There was no room for explanation from anyone, and more often than not, accusatory statements from me would cause others to become defensive.

I Can't Go On

One day, Gentry and I had just come home from the store and I looked for a reason to pick a fight. Suppressing my feelings was easy. Getting them out was not. For one reason or another, Gentry and I began fighting. I'm sure he was confused. What exactly were we fighting about? Standing in the garage, he said, "Renee, you have to do something. How much longer do we have to suffer for your grief?" I didn't even know what to say to that. I stomped into the house.

Over the course of the next few days, the fighting got worse and I called my mom. Things happen when you include your mom in your fights with your spouse. Mom and I ended up in an explosive fight and I just couldn't take it anymore. I couldn't take anything. I was at the end of my rope and I could see no hope in sight. Where was God? I thought He said He would never leave me or forsake me, and now He seemed more distant than ever. I thought He said He would be near to the brokenhearted. What does that even mean? To me, He was gone…He had bigger problems to deal with. Or perhaps I wasn't being spiritual enough? Perhaps, I had to fake it until I made it. Maybe I was supposed to "do" something in order to feel God closer to me, as if I had any control over what I was supposed to be feeling. Is God's nearness to us a feeling? Are we supposed to "feel" Him nearby?

What I know to be true about God is that He is near whether we feel him close or not. This is why He gives us His Word to rely on, not our emotions. His Word is truth and if He says He is near to me, then He is, regardless of whether I can feel Him. His Word is His promise to me that, "Even when you cannot feel me close to you, I am close to you because my word says I am, and I cannot lie." Numbers 23:19 and Isaiah 41:10 helped me see this even more clearly. These are the times I had to rely on the Word of God, not my feelings.

My younger brother, Daniel, had called to ask why I didn't call them more often to check on the family. I didn't have an answer. I was just trying to make it myself. I had a family of my own that I was trying to be there for. I had to be a wife and a mother to three children and I was also expected to be there for other family members. I couldn't. I just couldn't. My life was a complete disaster. In hindsight, perhaps Daniel did truly need me, but I didn't know how to be there for him.

I felt like I didn't have the right to grieve like my parents, or my sister in law…or my nephews whose father had died. I…am… the…sibling. What meaning did I have? That was a question I had to answer on my own. It's a question I had to find the answer to regardless of what anyone else thought or still thinks. My brother was and is still special to me. There are things he knew about me that no one else knew and vice versa. We were each other's secret keepers. The secrets that he took to his grave. The only one who held the special keys to my childhood and all my secrets was locked away in a gravesite at Arlington National Cemetery while my brother lives on in heaven.

Sam had a unique relationship with everyone. His relationship with his half-siblings was different than his relationship with me, but he always made each relationship feel 100% authentic and special. I could not relate to my much younger siblings in their loss journey any more than they could relate to mine. I didn't want to feel guilty for grieving when my parents, my sister-in-law and my nephews were also grieving. There were many times when I had wished it were me who had passed on. I even had thoughts that my parents may have felt that way, also.

I can't even rationalize now how I must have felt back then. I was not some well-put-together, confident person. I felt as if I had failed everyone in my life. I paced the living room when I was alone in the house. I was already broken from a life time of feeling inadequate and the one person who was always there for me was now gone. I placed my hands on the picture frames of my family…of my children. I knew I had become a lousy mother, so I thought. What I was really, was absent. I ran my fingers across their sweet faces and I looked into their eyes. They were so full of life, so full of love. I couldn't imagine my life without them; however, I was in so much pain, I couldn't fathom or think about how not existing anymore in their world could affect them. There was just nothing but brokenness and I didn't know how to fix it. I didn't know how to navigate through it either.

The house was quiet. I began to think of my life and ask myself how I could find meaning in it again. My relationship with God was pushed aside like every other relationship in my life. Not that those around me didn't matter, but I felt *I* didn't matter. I walked out into the garage and I grabbed a bottle of wine out of the fridge. It was one that had been there from the previous six months since I don't drink wine that often. I returned to my bedroom, walked to the medicine cabinet, opened it and grabbed my bottle of Xanax. It was the bottle my doctor gave me the day Sam was killed. The bottle was only missing the one pill I took so I could sleep after the 48 hour drive to Dover.

That experience seemed like a dream now. I hated taking medication. I hated not feeling in control, but really, I wasn't in control anyway. I was so out of control. I sat on the side of the bed, lifted my feet off the cold hardwood floors, and placed them on the bed frame. I could feel the shape of the wood frame as my toes curled around its edges. I held a bottle of wine in one hand and a bottle of Xanax in the other as I contemplated taking my own life. I could just go to sleep. *How much of each would I have to take for all the pain to disappear?*

I do know that there was this fear inside of me that didn't want to take that risk, but I just didn't know how to make the pain stop. I just sat there staring at each of these bottles. I felt like my life was over and as I lay back, tears streamed down my face. I cried my eyes out, and wished someone knew how to help me. Wasn't there someone who would pull me out of this hole? Wasn't there someone who understands how much pain I was feeling? I just curled up in a fetal position.

The thoughts began to cross my mind of my husband and children finding me dead on the floor. *Did I want them to experience the pain of losing me? Did I want them to feel how I was feeling in that moment?* So many thoughts raced through my mind. *What would happen to their lives if I took my own? How would my mother cope with losing another child?* Honestly, I didn't think much of my other family. I had never really felt like I belonged to them anyway. Maybe they would be relieved. Maybe all along they had wished it were me that was dead and not Sam.

All of a sudden, there was a knock on the front door.

What the hell? I thought to myself. I wasn't going to answer the door. I didn't care who it was. The text message came in from my best friend, Joy, who knew me so well. She had walked through many dark times with me. She had prayed with me and been there for me through two miscarriages, one that almost took my life. She had seen me broken before and somehow the Holy Spirit had told her to come to my front door. She called me and I didn't answer. She knew the code to the garage and let herself in. I could hear the garage door open to the house, the sound of her footsteps, louder and louder as she called my name and approached my bedroom door.

My heart was racing and I was paralyzed in my own fear and guilt and remorse and hopelessness. "Renee!" she called. She placed her hand on the doorknob and opened it. There I sat, staring at her. She stared back at me. She ran over and grabbed the bottles and I collapsed in tears. "What are you doing?"

I was crying so hard I couldn't even get the words out. Gentry came home soon after and I wondered what it would have been like if Joy hadn't come. What if I had succumbed to the pain? Yet, I was so relieved she came. I just needed someone to see my pain, just one person to acknowledge that I was not okay.

Things moved so quickly. My husband immediately got me help and I was placed on an antidepressant. I wasn't happy about that at all, as I again saw it as weakness. I don't see it that way now because it saved my life. After a few weeks, I began to think clearly again, enough so that I could function, but there was damage already done to my relationship with my mom. I didn't know how to fix it.

CHAPTER 10
THE OVERLOOKED GRIEF

You've been gone just over a year...I dream of you often as if you were never gone. This is a trial that has shaken us to the core, but I assure you, brother, I won't let you down. I am still amazed at how many lives you continue to touch. Christmas was not the same this year. Thank you for the laughter you left us and your amazing legacy. I miss you so much!! Until we see each other again. (Journal Entry -2013)

The pain, the loss, the grief...they all muddy the waters. Hearts are changed. People are changed. My cousin, Karen, would always say that misery loves company. I think there is some truth to that. Eventually, though, you have to make a choice to *not* live in misery. I didn't understand Robin's choices and so even we stopped speaking. It hasn't been time wasted. I couldn't self-reflect if I hadn't been on my own so to speak. I needed that time to heal...on my own.

I finally found the peace I needed. It's not that I didn't want reconciled relationships. What I do know is that, *"All things work together for good for those that love the Lord and are called according to His purpose,"* (Romans 8:28). Reconciliation does not mean the relationship goes back to the way it was. Reconciliation does mean that in the depths of our hearts, where only God and we can see, we wish them well. "As much as it can be done by us," we live in peace (Matt 5:23-24). I have chosen peace.

There was a time when I'd apologize for others wrongdoings just to remain in the relationship, but all that did was build walls around me causing resentment. I can't force people to like me or to be in relationship with me. I can't change people's hearts or make them forgive. My father always used to say and teach us to, "Take the harder right versus the easier wrong". The harder right sometimes meant letting go of others in my life, for them to find their own way…wherever God leads them.

It means confronting the dark places in our hearts that we don't want to face. It's realizing we all are not perfect no matter how much we want the world to think we are. We all have experienced brokenness at one time or another. We all fail. It is the reason I am so thankful for Jesus' redemption. I have had to apologize for my areas of wrongdoing, even when others have not. I have learned that in the taking responsibility for my part, God has allowed me to see Him *change* my heart.

On the road to forgiveness, the grieving still continues. Grieving continues in many different ways and is unique to every individual and situation. This is not commonly known though. In fact, Elizabeth Kubler-Ross's "5 Stages of Death and Dying" has now become a frequently used coined term for all grieving people. While there are "stages" a grieving person must go through, in my experience, the "Five Stages of Grief" can often be a misconception in the way the general public expects people to grieve. I am not the only one who thinks this. Elizabeth Kubler-Ross never designed it as an absolute on grieving.

> "In her book, she lists the five stages of grief that she saw terminally ill patients experience in the face of their own impending deaths: denial, anger, bargaining, depression, and acceptance. However, she never intended for her five stages to be applied to all grief or to be interpreted as a rigid, linear sequence to be followed by all mourners," (Carroll and Wolfelt, Ph.D.; 2015).

Friends and family wait for the stages to pass until you are yourself again. Your normal self. Right? Wrong. I was asked frequently what stage of grief I was in. When other families had experienced a death, I would overhear well-meaning friends say, "They must be in such and such stage of grief." I think of grief as a journey, more so than stages. You never know what emotion you may be feeling at any particular moment.

Grief lasts a lifetime. Does love come to an end? Of course not. Neither does grief. I know other Gold Star Families who are still angry ten years later. Did they skip a few stages? Are they refusing "acceptance"? It doesn't happen that way. Grief is messy. Grief for a sibling? It's awful. You feel you are almost assigned the task of making sure your parents are okay and the widow is okay. Society says, "The sibling carries the load for the ones who should really be grieving". It's lonely and it sucks!

The phone rang. I looked at the caller ID. It was Aunt Mabel calling again for the third time in a row. I had no desire to talk to her. *Great, a voicemail.* I put the phone up to my ear. "Renee, this is Auntie Mabel. Why don't you ever answer your phone? Look, if you don't want to be a part of this family anymore, then fine. If you don't have the decency to answer your phone and be there when we need you, then fine." Oh God. I just can't deal with this old woman. She has to make a dramatic production out of everything. What the hell does she mean, "If I don't want to be a part of this family"? She always does this. She always threatens my place in the family and then plays the victim. I'm just done with her!

Yes, sometimes people have hurt me for no reason. Perhaps they have reason and we all know that, "Hurting people hurt people". I used to carry so much anger towards others, but once I began searching God's view on forgiveness, I began to understand. I feel sorry for all who carry the weight of unforgiveness, bitterness, and resentment, because I know the weight of that curse. I know it's a heavy cross to bear and I know Sam would not have wanted things to fall apart in our family like they had.

Through this process of forgiveness, God taught me that when there are two sides battling, the one who meddles is the one who gets most of the direct fire, and I surely did. There was a time, admittedly, that I wanted to be right and I wanted justice in my life for the places I felt God did not defend me. All along, He was teaching me about what it was to walk in forgiveness and love others, regardless of whether or not they were ever sorry.

It is a hard road to self-reflection as you search your own heart when you feel hurt by others. After Sam's death, there was a period of about two years that I was in complete and total emotional breakdown. The first year was a blur. I was in shock. Anyone who tells you the shock wears off after a few weeks is either lying or has never experienced the death of a close loved one. Shock lasts a long time. My actions became mechanical. It's commonly known as "going through the motions," but my mind was numb.

The second year, the fog lifted and I was keenly aware of my new life. At the same time, I didn't really understand how to navigate through the grief inside this new life. Unfortunately, it was within those first two years, when my family dissolved, secrets were revealed, hurts surfaced, and no one knew how to deal with any of it. I couldn't even deal with my own grief let alone all the pressing issues of family dynamics. Our once happy family was a distant memory and a constant reminder of what could have been. It was not just my family. It has happened in most families who have experienced such trauma. It is no wonder why so many families fall apart.

Often, more so than not, people won't be sorry for hurting you. It takes a person who truly wants to change their own heart to choose to self-reflect. That is painful. Sometimes pain is the only way we learn, the only way we learn to change ourselves. We can't always focus on changing others because that's impossible. I can't explain why others make the choices they make and I can't explain why people choose to retreat in times of trauma and grief. I can't explain why I did, but I do know that God will use everything that happens in our lives as forward movement toward a greater purpose.

There are ways of setting healthy boundaries with people who hurt you. This was a process I had to learn. For me, I had to disassociate myself from Aunt Mabel. Unfortunately, that meant some other family members also. I couldn't allow her to hurt me anymore and heal at the same time. You must learn how to allow people to treat you. You must learn to deal with them in a manner which expresses love and forgiveness, and sometimes that means breaking off the relationship altogether. This is where the rubber means the road. The cold hard reality of not having the perfect family is revealed and you learn to let go and be content with that. Even though she passed away several years after Sam died, I still had to let go of the hurt.

Most of the time, it's the enemy of our heart who wants to cause division. If he can distort circumstances, if he can cause a wedge and division in a relationship, seemingly he has won the battle. But truly, when you look on the inside and you see how your role has affected other people, you may find your battle wounds. You may be bruised and beaten, but you haven't lost the war. And even though my brother died defending his country, he didn't lose the war, because he never gave up fighting.

While my brother also had little tolerance for injustice, he had a way of addressing things in love, or he was willing to over-look the incidences for the sake of the relationship. He could correct me in a way that I could hear him, learn from what he was saying, and change my mind. He could be stern with me in a way that I understood he was coming from a heart of love and restoration. I, on the other hand, have had to learn through some lost relationships over the past 20 years, that I cannot always retreat when someone causes me pain or it will be a very lonely road. You cannot influence people if you are constantly pushing people away. You have to forgive. Period.

It wasn't just the reality of changing family dynamics that I had to face. It was the reality that Sam was, in fact, dead. He wasn't going to come home from deployment. He wasn't going to call or text some funny story. He wasn't going to smooth things over within the family like he'd always done before. He was gone.

When Sam died, I realized I didn't know who I was without him. I spent a lifetime living in his shadow and I was comfortable there. I was the little sister. I was "the girl". I had nothing in common with our father except my eyes. Sam, on the other hand, knew he was destined for two things: to be a fighter pilot and to build Mustangs with Dad. Everything Sam learned about Mustangs, he learned from Dad. That left little room for me. As I self-reflect, I realize that I was never jealous of my brother and his accomplishments. The only thing I had wished, was to know my dad was proud of me for being me the way he was proud of Sam for everything he was.

If we allow it, resentment can build after a military death, especially between children and parents. My brother died a hero's death. How on earth does one compete with that? I was no longer Renee, Sam's little sister. Now I was Renee, the sister of a Marine who sacrificed his life for his country. The highest honor one receives from such a great sacrifice was an honor I never wanted. I didn't want to be a Gold Star sibling…who would? I was angry Sam left us. I wasn't angry at him, I was angry he was gone from our lives. I wasn't comfortable being the oldest sibling now. I didn't want to be responsible for my other younger half-siblings and take on a role that only Sam could play. Maybe I placed the expectation on myself that was just too painful. When my then sixteen-year-old half-brother Daniel called me to ask why I never called them to ask how they were, I knew. I knew they expected something I couldn't give them and I backed away even further.

I was the sister, the wife, and the mother who helped others succeed. I loved helping others reach their dreams and their goals. I loved to see other people successful and pushed them to strive for more. I loved being the person behind the curtain who made everything run smoothly. It was safe. I could hide. I could be at ease in the fact that I was not set up to fail. There was no room for failure in my family anyway. So, I never tried.

I loved watching my brother succeed and I knew that he was *my* brother. "That's my brother," I would brag. I was so proud of him. I was content living in his shadow. It took no risk. I didn't have to fail and he was so good at succeeding. I was safe under his wing.

Now he was gone. What now? Who am I now? I spent the majority of my life hating myself and never feeling like I quite measured up. I started school a month after Sam died. I wanted to be a lawyer or work in government. I had a passion for justice. I later decided I would switch to a business degree in finance. I wanted to be my own boss, perhaps an entrepreneur. That didn't work for me either. I tried to discover who I was through some degree or profession. I was not content being "just a mom". It wasn't good enough for me and I knew it certainly wasn't good enough for my father, or at least I felt that way. I felt I was now living in the impossibility of measuring up to the shadow of my brother's ghost.

I gave up trying. I pushed my husband to get his MBA and another Bachelor's degree. I pushed my children into their dreams, or what I thought were their dreams. I never wanted them to resent me for pouring all my energy and belief into one child. I spent so much time trying to not be my father and stepmother that I forgot exactly what the point was. I got lost and I grew lonelier. I felt more and more insignificant. I wanted to be anywhere but where I was…where I was in the physical and where I was in the emotional. I just couldn't cope with losing myself to my own expectations.

I walked around, completely lost with no direction. How do you navigate a ship with no compass? God used to be my compass, but He was as silent to me as the ship is to the sea, lost in the vastness but still yearning to find direction. I needed direction. I needed courage. I needed to know it was okay to want something bigger than myself. I wanted to know it was okay to matter for just being…me.

I continued to search for answers only God could give me. I needed answers as to how I would go on. I was now an only living child to my mother. I had zero relationships with my half-siblings and my father. Sam's death changed everything. I began to face the realization that I would be the only one to care for my mother in her elder years. I say this not intending that it will be a chore. I couldn't imagine not sharing life or growing old with Sam. Who would laugh at my mother's endless collection of…collections? We would playfully argue about who gets these items one day. We knew neither of us wanted her endless supply of glassware, but also knew that the only thing that really mattered was to make the best of the times we all shared with each other.

All of that possibility and all of those future experiences were gone. When I thought I now have nothing to share with my big brother, it grieved and pained me so greatly. I would have given anything to have him there with me. I'd give anything to sit back in the chair beside him and hear my parents boast about his life and what a wonderful person he was as they always had before. He was and always will be, my hero. *How could my parents ever be proud of me like they were of Sam?* Siblings can often feel this burden when part of them has died, particularly during military service. Pride is a powerful thing.

I searched my brain for my mission to honor my brother. My heart, on the other hand, was stuck. My heart wanted a life of my own, separate from being a Gold Star sister, and I did not know what that looked like. I spent an entire life with a person who knew everything about me and I completely lost who I was when he was gone. I couldn't make sense of my life, and I felt as if I was going nowhere.

Feeling Hopeless as a Parent

My biggest role in life was being a mother. I knew that. I had three precious children to take care of and I was struggling...greatly. My marriage was falling apart as was my relationship with my oldest daughter, Kylee. I was completely lost in my role as a person, as a sister, a daughter. How could I parent a teenager who had also experienced a trauma?

Kylee, just a twelve-year-old little girl, had to help pick up her grieving mother off the floor. Somehow, I expected her to have better grades, a better attitude and stop showing me disrespect. I demanded obedience from her when the lines between parent and child were jumbled. The fights got worse and I could feel her anger and hate towards me. Not because she truly hated me, but she hated that I was emotionally absent. I was so caught up in my own emotions and grief experience that I felt the loss was only mine. I couldn't help her any more than I could help myself. I truly expected my children could dust themselves off and move on. She too, was all alone.

Years later, my daughter, Leah, who was five at the time of Sam's death, still cries for him. Usually around the Christmas season, we all start to get a little on edge. I tend to forget that these little children have seen more pain, trauma, and separation in family than most children their age ever have to. Of course, there are going to be suppressed feelings that have to get out.

“Kylee, what is going on with you?” I’d ask her.

“Nothing Mom,” she answered as she rolled those fourteen-year-old eyes.

“I’m asking you a question, do not disrespect me!”

“LEAVE ME ALONE!” Her volume escalated.

So did mine. “I will NOT have you speaking to me like that!” I shouted.

“You always do this! I don’t want to talk, just leave me alone!” Kylee replied in frustration.

I continued to push, “Young lady, you will tell me what is wrong with you!” As if there were something *wrong* with HER. She didn’t know what was wrong any more than I did.

“I hate my life! I hate being here! Everyone at school hates me and I don’t fit in anywhere.” She began to cry. “Now I have to deal with all of this.”

All of this? What did she mean by that? I stood there shocked. She hated her life. I knew she was not talking about the usual teenager comment, “I hate my life because I don’t have a new iPhone,” but she hated her life.

“Mom, I’m sick all the time. I have to take shots every time I eat something. Do you know how embarrassing that is? I quit ballet because I didn’t want to do anything after Uncle Sammy died. Now, I’m not just the girl with a disease, but now I’m the girl whose uncle died in war. And now I can’t even talk to several of my family members. Our family is so screwed up! WHAT DO YOU THINK IS WRONG WITH ME? WHERE THE HELL IS GOD? I JUST

WANT TO BE NORMAL!" Kylee cried.

Well, okay. I didn't expect all that. I didn't know what to do with all that information. I didn't know how to answer her. I couldn't fix any of it. I had zero control over the trauma in her life and her process of emotions and grief. I certainly couldn't explain to her why God allowed certain things to happen. I assumed she was a rebellious teenager, but what she really was, was a hurting little girl. I was her mom and I couldn't make things better. I couldn't "fix" her.

What could I do? I had no healthy experience at dealing with a hurting teenage girl who was living with trauma. I fought everything within me to not be emotionally disconnected like my father and stepmother, and to actually allow my child to express her feelings. While that is ideal, it didn't really work out the way I planned.

I began to make wrong assumptions about Kylee. Others did too. Children will act out for deeper issues and reasons. I did. Others do too. I learned that the greatest gift a parent can give their struggling child is to take the time to dig deeper and understand from their perspective what is happening in their little hearts. I didn't know this. I just wanted our family to be whole. But we weren't. We were far from it. I was trying to pretend we had it all together just as much as my father always had. When you stick your head in the sand, it is only effective if your goal is to suffocate yourself or give others a big heaping view of your rear-end. The fact is, it took more courage to admit I was broken than it did to live a lie and allow my children to experience the aftermath, picking up their parents' mess. This holds true today, whether the child is fourteen or thirty-five.

The bubble I tried to place around my children burst the day their uncle died. I had no way of protecting them. I wasn't equipped to grieve and shield them from the pain at the same time. I couldn't protect them from the actions other people took against their innocent souls. I had to learn that it's okay to not have all the answers as a parent. It's okay when our kids screw up and we don't know how to make it right. I didn't know that then. I felt like a complete failure. One thing is for sure, my children are kinder people because of the beauty that they developed from their difficulties and grief experiences. They are the most forgiving little people I've ever known. Regardless of the pain I or others caused them, they have continued to pray for us all.

Sometime later, I opened up the computer one night to look through pictures of my son between the ages of one and three. Kylee came in. "Mom, what are you doing?" Kylee asked quietly. The little ones were all asleep.

"I'm going through old pictures of little Ethan" I replied. "Kylee, where was this picture taken?" Kylee began to tell me story after story of little Ethan's life. He was just fourteen months old when Sam died. "I can't remember any of this. I missed out on so much." I said to Kylee.

She replied, "Mom, you weren't there."

"I was too there. I took the pictures."

"No, Mom, you were *there,* but you weren't *there*. We lived for two years without a mother."

I began to cry. I looked into her eyes. "Kylee, I'm so sorry." I grabbed her and hugged her.

"It's okay, Mom. We'll be okay."

Kylee was and is wise beyond her years, and fights to find her own place in the world. She is still coping with her own struggles from the rejection I caused during that season, and the death of her uncle. She wouldn't be the amazing, kind, fun-loving girl she is now if she hadn't experienced real life and real pain.

Feeling Forgotten

It's an indescribable pain to lose someone you grew up with. I spent more time with Sam from birth to 18 than anyone had, even my parents. I write this, not to minimize their loss, but rather to express mine. Each one of us, regardless of the loss, but specifically siblings, need our grief validated. We need to feel like our grief matters. We are desperate for our pain to be acknowledged just as much as our parents, spouse, and children. No one's grief is greater than another, it is individual. It is a unique journey. Siblings are often robbed of feeling like they can grieve freely.

I don't think one single person ever asked me what it felt like to be me, to have a sibling that died, unless of course they had experienced a sibling death themselves. This is not only an injustice to ourselves, but to others as well. There is this expectation placed on us to be the caretakers for everyone else. I did have close friends who were truly genuine and were caring and thoughtful. And there were my cousins Karen and Pam, who saw the worst of me at times and loved me through it all. Even though they were there for me, I had to and still must walk my grief walk in my own unique way.

Most grieving people don't really know how to receive "words of comfort", nor do people really know what to say most of the time. Well-meaning people say some of the most hurtful things after a loss. I've heard things like, "My dog died, I know just how you feel." Or "Great Aunt Sarah died when I was six, I know how you feel." *Lord, help me keep my mouth shut!* Unless you are the one grieving, you really do not know how they feel. I think the most hurtful statement

I heard was right after Sam was killed, when a person told my mother, "Sam would rather die than come home injured." These words were meant to bring comfort?

Here's a great rule for anyone who wants to comfort a grieving friend or family member: Don't say anything that begins with "I know". You don't *know* anything. The most helpful words I ever heard were, "I'm sorry for your loss". My loss was acknowledged and I knew the person was genuine in their sympathy. Please do not bring up anything about "Your brother is an angel now," or, "I guess God needed another angel". I especially hated, "Sam served his purpose here," and, "It was his time" or, "He died doing what he loved". *Really? He loved being in war and shooting terrorists?* Don't tell me that! What he loved was his family, his wife, his sons. He loved his Marines too.

So, I've received comments from both ends…the helpful and well-meaning, and the "I want to punch you in the face for that". Truthfully though, there are really no words of comfort that will help a grieving person. All you can do is *show* them you care.

Sometimes, when people just called me or wanted to be with me brought the most healing. I didn't want to feel excluded or that I had the grief disease. I wanted people to treat me the same as if Sam had never died, but things had changed and relationships had changed. I withdrew just as much as others did. People said, "Call if you need anything". I wouldn't call. Grieving people don't call and ask for help. Most days, we can barely get out of bed or feed ourselves. We are not going to call you and tell you how crappy our day was because we don't know how we will go on without our loved one, even years later.

After losing someone as close as a sibling, you are literally a different person. You try to be the person you used to be, but you can't…at least I couldn't. This can cause feelings of guilt. I felt guilty for the inability to be the person I once was. I felt guilty for missing him so badly, as if missing him was connected to my inability to move forward. Many years later, I still miss him. I miss him even more than I missed him the day he died. The more time that passes when he isn't here to enjoy life with me, the more time I go on missing him and the future I dreamed of having with him and our families.

I suppose it is self-inflicted guilt which caused my discomfort when talking about Sam. I know, I felt it. I still do at times. There is nothing wrong with it. It just is what it is. We all have to grieve in our own way. Acknowledging one's grief is the best gift you can give that person. Ask me about Sam. Let me tell you stories about him. He loved life and he loved making people laugh. Let me tell you stories that will keep his memory, his legacy, and his laughter alive.

CHAPTER 11
WHERE I GO, THERE YOU ARE (2014)

Leaving our home

God, where are you? Why do you seem so silent? Is there any purpose in any of this? Do you have any purpose for me? Who am I? These are questions I frequently asked myself. My life was chaos and I didn't know how to navigate through it. Sam was gone and I was estranged from my father and siblings. My relationship with my teenage daughter was crumbing right before my eyes. My marriage was strained. Gentry didn't know how to help me. I no longer wanted to communicate about anything.

I needed a new start. There came a point where I saw the crossroads approach. I longed for a better tomorrow, something to look forward to. Thankfully, my husband and children brought me a certain measure of joy that I needed to keep moving forward, even within the messiness and conflict.

I noticed my attitude getting bad simply because I was getting impatient with God. Not only was grief lonely, but it was even lonelier when I thought I no longer had anything in common with those who I was in relation with. Eighteen months after his death and I felt as if I must fake a smile and say, "Yeah, I'm fine. I'm happy," but I was not. Every morning that I woke up reminded me that our lives would never be the same. Sitting around the dinner table on Sam's birthday with his wife, sons, my mom and stepdad, only reminded me of what once was.

I yearned for a new start where people didn't have the expectation of a better version of me. I wanted to be in a place where I didn't have to excuse who I was because I was different than I was 18 months ago. I was and am changed, changed forever. Connections were much more difficult no matter how hard I tried. This was compounded when people obviously thought, "You should be healed by now" or when they would try to pacify me and pretend because they were uncomfortable with my grief. Eventually, there came a point when people moved on and forgot how much pain I was still in, so I learned to withdraw. It became easier conversing on social media than to a real face. My heart had already left my hometown of 10 years, and the rest of me just waited to follow.

Gentry applied for a job in Fort Worth and the thought of moving to a new home and city was a welcome relief. I couldn't wait to move somewhere, anywhere where I could just be me, somewhere where I could tell my story only if I wanted to. I swore I wouldn't "friend" anyone new on social media in a new town, so they wouldn't "know" anything about my story. I could pretend to be someone different. Even *that* promised I would be extremely lonely. Start talking about how your brother was killed in war to a new friend, and you get the glossed over death stare. The "What the hell do I say to that?" look.

So, what happens when you can't find the right words to say to someone who is grieving? There is a dichotomy that happens. Most of the time, I don't even know what to say back. On one hand, you're angry because people aren't saying the right thing. You're angry because they don't know or say what you need to hear, even if you don't know what that is. On the other hand, you can become angry for people not making any effort to say anything at all, even if it is the wrong thing. Then there is the awkwardness until you eventually stop opening up, you put on a façade, and you deal with your own pain… alone. I was angry when some people in my life pulled away, yet relieved at the same time. I would be angry I wasn't being pursued, yet relieved when I could be left alone. Emotions fluctuated all over the place, sometimes moment to moment. Even so, I couldn't wait to relocate to a new place.

The family all piled in the car for one last sunset on Destin Beach which had been our home for ten years. It was the end of spring and the summer heat was quickly approaching, but the evenings were a cool 78°. We pulled along Highway 98. The girls were in their sundresses ready to play in the sand. Little Ethan was in his bright blue glasses that magnified his big blue eyes. He was eager to get out of his car seat and hit the ground running. I got out of the van that one last time. The ocean breeze blew through our hair as if it knew we had nowhere to go. The children immediately headed for the water. I warned them not to get in past their ankles even though I knew they wouldn't listen and would get themselves soaked. They began splashing each other playfully as I stood back to observe. Gentry approached me and put his arm around me. We just stood there together admiring the beauty and the vastness of the ocean.

Kylee ran over and grabbed her dad to come and play. Gentry joined them as I stood there alone. I began to walk. The beach was empty as if we were the only ones who wanted to enjoy the beautiful sunset. I was several yards away and decided to take off my sandals and allow my feet to sink down into the white sugar sand that Destin is famous for...one last time.

I stood at the ocean's edge as the orange sun reached the horizon. The sky was full of pink and purple hues. As I stood there, the cool water crashed over my feet which were now sinking, completely covered in sand. I reminisced about everything I would miss about this place. This city was where we brought two of our babies home.

I began to the think about the faithfulness of the Lord and how much I'd had to wholly trust in Him to see us through such difficult times. As I stood there, I wondered about our future, where would He take us, and if we would be happy. A song played in my mind, *Oceans* from Hillsong.

"You call me out upon the waters,
The great unknown where feet may fail.
And there I find You in the mystery,
In oceans deep, my faith will stand.

And I will call upon Your name,
And keep my eyes above the waves.
When oceans rise, my soul will rest in Your embrace,
For I am Yours and You are mine" (Hillsong United, 2013).

I longed for a change, a different perspective, a place where we could heal. I didn't know what that looked like, but I knew our season in Florida was up. God was directing our steps. I had no choice but to trust Him. My eyes focused on Him and only Him. I couldn't make it on my own. I knew it. He knew it.

I looked back at the kids playing with their dad. I couldn't help but think of my nephews living without their father. I remembered what it was like to be a child, carefree and madly in love with my own dad, my superhero, and realized that those days were long gone. I silently prayed to God that we were doing the right thing and asked for a sign that we would be okay.

Just then, two red and white planes entered the sky. They were T-34C Mentors. My brother had flown them in Pensacola when he was training to be a F-18 fighter pilot as a strapping young Marine ready to take on the world. I stopped and shaded my eyes with my hands so I could get a better look. I knew right away what they were. I have a photo of my brother standing in front of one of those planes.

They began to fly in separate directions as if they were putting on a show for us. The sound of the planes reminded me of how fortunate we were to live near a military base, our community, and our family. They began to fly back toward each other. They sped through the air and crossed each other, the tail smoke followed behind. I could feel the tears well up in my eyes as I thought of Sam, so young when he was in training with a full life still ahead of him. I remembered how I took every opportunity to brag about him. The planes turned upward towards the colorful sky about to cross each other again. I stood there in awe. I couldn't believe it! They made a heart over me. Tears flowed down my face as if Sam was telling me he was there with me. He loved me. We would be alright.

It's as if God opened the heavens for me in that exact moment, the exact moment I needed to hear from Him. I needed something to help move me forward. This was one of those moments when I felt like God kissed me and allowed me to step into something supernatural. A moment that could have been missed if I wasn't exactly where I was supposed to be at that exact moment in time.

I looked over at my family once again. They were giggling and pointing up in the sky. Their laughter filled the air, along with jet noise in the background. I knew this was a moment that would stay with me forever. Somehow, I knew we would be okay.

A New Start

Our new start in a new place with a new home and a new job was perfect. The move came with a hefty promotion and it was everything Gentry and I had worked for, for over a decade. It was time for us to experience everything as a first. We'd have to find a new church and a new home. We felt free. We could go and do anything we wanted. The possibilities were endless.

Everything seemed to be falling into place. Nobody knew me. Nobody knew our family. Nobody knew I was grief-stricken. On the exterior, we appeared to be perfect. I was familiar with this perfect. I grew up in this perfect. I had just obtained everything I ever wanted, or thought I did.

I had absolute control of everything around me and I liked it. We had spent years living in a modest home, working with everything we had inside us to become "something." And here we were. We had arrived. I was naive enough to believe that my grief wouldn't follow. I believed that changing locations in the physical would somehow surgically heal my heart and my mind. It's amazing the lies we tell ourselves.

Sometimes change is necessary for us to get through the trauma and the grief. There is something about leaving the old behind that helps us move forward. While I had wonderful memories of my home back in Florida, there was a constant pain that reminded me that it was the place where I found out my brother had been killed. Every time I looked at the clock on the counter, glanced at the nightstand where my phone was, or my mom happened to call me before 9am, memories triggered the PTSD. My mom actually stopped calling me in the morning because every time I answered, I asked her if everything was okay. Now we talk mid-afternoon. Everywhere I looked or went reminded me.

Mom wasn't home the morning two Marines showed up at her home. I am thankful they had missed her by minutes. She had left for work already. Her home, a place of warmth and love, is a place where we spent many a day over the past twenty-five years, as children, and then with our own families. I couldn't imagine the pain associating her beautiful home with the place Mom found out Sam had been killed. It was our home. Mom's house was always "home" and, thankfully, we still have that as a family.

Change, when the timing is right, can be a very good thing. Perhaps I was a little bit too optimistic and a little too desperate. Change meant leaving that place of trauma and starting over. This was the perfect opportunity for us...to change our exterior circumstances, even though the inside of us was still very much broken. We tried several new churches over the next nine months. We didn't want to jump into anything too quickly. At the same time, our dream home was being built. Gentry was finishing up with his Master's. Kylee started at the new high school. Leah jumped right into gymnastics as a competitive gymnast and little Ethan was in soccer. Everything seemed perfect.

We searched for a new normal, and life began to happen for us. Everything was exciting because everything was new. We now lived in a major metropolitan city that was far more exciting than our small town back in Florida. Fort Worth was the perfect in-between of city life and country life. The children had endless opportunities and so did Gentry and I. The people were genuine and we felt like we were living in a small town with a lot to do and a lot to eat. Not a single soul knew anything about us and that's the way I wanted it. Gentry transferred units in the Air Force Reserve. Things were great and I felt like we were moving forward and making progress. I knew this was the right move for our family and I loved life in Fort Worth. I'll never forget the first time we ran into another person we knew. It's amazing how quickly that happens.

Life couldn't get any better.

Maggie

I smile when I hear her name. We were at a company picnic for Gentry's squadron. It was a beautiful October afternoon. In Texas, that usually meant it was 95 degrees outside. Coincidentally, we ran into a couple we had been neighbors with when we were stationed in Phoenix, AZ, over ten years prior. The military world truly is small, which makes the Gold Star world even smaller.

"Gentry, I'm not feeling well. I think I need to go home. I think I'm coming down with the flu," I stated.

"Okay, do you want to leave the kids so you can go home and rest?" he asked, always willing to help take a load off me. He was and still is always amazing like that.

"I'd love that," I said. I went home and crawled into bed. The rest of them came home just a short while later.

Gentry came in and asked how I was. "I bet you are pregnant."

I literally laughed. "I'm not pregnant. You know how hard it is for me to get pregnant, we tried every month for three years to get Little E.

Plus it's just not possible…and I'm almost 40," I explained.

Gentry smirked. "I knocked you up," as he did a little dance and made me laugh in my flu-like misery.

Later, close to bedtime, Gentry said again, "Maybe you should take a test," with a grin.

"OH. MY. GOSH! Would you please shut up about this? I'm not even due for a few more days anyway," I said jokingly. I knew he'd continue joking with me until I proved to him otherwise. I was 100% certain I was going to waste money, which I hated to do, but in this instance, was necessary to prove him wrong. "That's it, I'm going to CVS." I said. I ran to the drug store and picked up a few pregnancy tests and some flu relief medicine. You can never be too sure. Now at home, I sat on the toilet, took the test, and looked down. There it was. Two BOLD solid pink lines staring back at me. I began to cry.

Gentry walked in. "Why are you crying on the toilet?"

"Because I'm pregnant. I'm sooooo oooold!" I said.

Gentry laughed. "My hot pregnant wife! I knew it! I told you so!"

I didn't know whether to be thrilled, terrified, or sad. *What were we going to do now? Our fourth child. I always wanted six children. I had had three dreams that I would have twins. Oh, my gosh. What if it's twins!* I thought. Now what? *This has to be twins because there is no way I am getting pregnant again after this one.*

A couple months passed and we got the ultrasound results which confirmed it was a singleton, and a girl. The children were excited except for Little Ethan who stomped his feet because he wanted a baby brother. I told him, "You'll have to take that up with God." Not only was it a girl, but it wasn't twins. *So, what was up with all those twin dreams? God, please don't let this mean I'm going to get pregnant again!*

"What are we going to name her?" I asked Gentry.

"Well, I picked the last one. So, you get to pick this one."

"How about we name her Sara Margaret? Sara after your mom and Margaret after my Nanny." He loved it. "We can call her Maggie. I always wanted a Maggie," I told him.

It wasn't too long before I began to struggle. While the external circumstances were great, the anxiety began to creep in. Pregnancies were hard on me physically, and this time I was afraid of feeling joy… that real joy only a new baby can bring. The joy that makes you want to burst every time you smell that little forehead and kiss those tiny little toes. That kind of joy I hadn't felt in a long time, since before Sam had been killed. *What if God pulls the rug from under me again? What if there is something wrong with her? What if there is something wrong with me?*

"Gentry, I forgot something until now," I told him. "Many years ago, when we almost named Leah, Sara, we had a change of heart. I remember sitting with my sister and talking about possibly naming another girl Sara. She had told me how she was furious because she wanted the name Sara for her daughter one day, but then realized how ridiculous it was to be angry over a name. So, she told me that she wanted me to name our next child Sara if we had any more. But, I couldn't take that away from her. I promised her that she could name her daughter Sara if she ever had one. So, we can't name our little girl Sara, because I made a promise to Kelsey."

Gentry understood. Ann is the middle name of both of our mothers. "How about Margaret-Ann Hope? Because she is our new hope for our future, hope we've been needing," I said.

"I love it," Gentry said. My ninety-five-year-old Nanny passed away soon after we decided on the name change. I knew it was meant to be.

The Birth

"Weeping may tarry for the night,
but joy comes with the morning."
(Psalm 30:5b; ESV)

With each passing day, I grew more and more terrified of the birth. I was facing blood transfusions if I ended up with another C-section. I was severely anemic and needed iron. Six weeks before the birth, I began I.V. iron therapy, praying it was enough time to bring my iron stores up. I told my doctor I didn't want another C-section. I wanted to have her naturally. I was aware of the risks, but I also knew a natural birth was still a much lesser risk than having major abdominal surgery and possibly bleeding to death.

The morning of my scheduled C-Section came. I went to see my doctor. "Doc, I can't do it. Please, give me more time."

Dr. Damien said, "Ok, I'm giving you one more week. If you're not in labor spontaneously by next Wednesday, I will have to do a C-section. I also must warn you, if you go into labor I cannot guarantee that the doctor on call will allow you to have a vaginal birth after your previous C-section, even though you successfully had one with your last child."

"Ok, now what do I do?"

"Well, you can go to a Chiropractor, or have lots of sex, or both," she smiled.

"Um, I think I'll try the Chiropractor. There is no way this body is attempting anything else," I replied as she chuckled at me.

I rushed home, found a prenatal chiropractor and made a 7 p.m. appointment. "I'm just going to adjust your hips a bit to make it easier for the baby to get through."

I thought, *Whatever, Lady. Do what works.*

I went home and straight to bed. I woke up suddenly at 1:30. *Oh my gosh, what the hell was that? That chiropractic adjustment couldn't have worked this quickly.* I went to the bathroom as another strong contraction began. *I should probably be timing these. Ok, five minutes. I'll go lay back down. Wait, these are getting closer and stronger very fast.*

"Gentry, I think I'm in labor…GENTRY! Gentry wake up! I'm in labor!"

He rolled over away from me, "Ok, honey." He went back into his snoring state.

Ok, he's useless right now. As I walked back to the bathroom, I felt more and more pressure. *Time check. Oh my gosh! They're two minutes apart.*

I went back to the bed, grabbed my pillow and hit Gentry over the head with it. "GENTRY! I'M IN LABOR! GET-UP!"

"I'm up, I'm up. Wait. What? What's happening?" he said in a daze.

"My contractions are two minutes apart. I have to get to the hospital or we are having a baby in the car!" I yelled in pain. "Go get the kids!"

The kids loaded into the car and we headed to the hospital. I had no idea what doctor was on call and I was facing some potentially serious complications. I was terrified. Gentry drove as fast as he could and my claws were now imprinted into the door handle. There was something else happening within me. With each contraction, I felt the pain of losing Sam.

When we reached the hospital, they quickly got me into a room. Fortunately, there was a pull-out sofa and all three kids went back to sleep. Thank God for small miracles. It was about 2:15 a.m. and the only thing I could think of was to put on my new Hillsong album for some peace and ask for some drugs. "How quickly can you get me some drugs and an epidural?"

Soon after, I begin to fall asleep. I did not experience this with my son's birth. Epidurals are there to assist in pain management, but I felt nothing. I realized it slowed my labor, but this was nuts. Not only did I sleep my entire labor, but my children did too. Everything was so peaceful. Peace which I still can't explain. Supernatural peace filled that delivery room during those early morning hours. I woke up only to realize that my doctor was on call that morning.

My labor slowed just enough for my doctor to arrive and deliver Maggie. Just as she was about to make her entrance, a song began to play. Two pushes and she was out. I'll never forget that moment, now known as Maggie's song.

"Love laid its breath against my chest,
My skin was thick but you,
Breathed down all my walls.
Oh love, like the fire steals the cold,
The ice wore thin as Your,
Light tore through my door,
You have my heart."
(Hillsong United; 2015)

It was the most beautiful moment when pain and joy collided. I remember crying as I lifted that baby girl to my chest. This beautiful gift God gave to me. He allowed me to birth her, He allowed me to experience this joy after all the pain. I knew everything was going to be okay. I knew God's Spirit was there with me.

"Renee, you did a great job. You are a walking testimony for so many women. You advocated for yourself…it saved your life. You saved yourself from a surgery and blood transfusion. You should be so proud of what you did," Dr. Damien told me.

I lay there, not really thinking I did anything any other mother wouldn't do. The doctor was bursting with smiles, "Renee, I wasn't sure you could do it. I doubted you, but, you knew you could do it. That's what makes this so incredible. You did this! You need to be speaking to other women about this. About advocating for yourself. I'm so proud of you."

She was proud of me? A woman who hardly knew me was proud of me? She's delivered 3500 babies and she told me she was proud of *ME.* I wasn't used to hearing those words and I didn't know what to do with them. I just held my baby girl in my arms, ever so thankful to God for guiding me, speaking to me, and delivering me…in more ways than one. I felt Sam there with me in my heart, cheering me on. I knew he was proud of me too, for taking an impossible step forward. Something bigger than myself happened that day. It was time for me to start believing for impossible things again.

2015 - Maggie's newborn picture taken on Sam's ABU shirt and the American flag he flew in his F-18 over Iraq during Operation Iraqi Freedom. I had that flag for twelve years and this was the first time it came out of its original packaging.
Courtesy of Life in Design Photography.

CHAPTER 12
JOURNEY TO HEALING

Meeting the Horses

The road wound through the old country town of Midlothian, Texas. It was early March and the weather was just beginning to get warmer. You could see new life coming forth from the tree branches that shaded the roads. Rays of sunshine shone through the trees as we drove down the two-lane country road. We'd been driving almost an hour and I wondered how much further we would have to go before we finally found a beautiful white fence lining the road.

After years of various traditional therapy methods that seemed to get us nowhere, I came across a method of therapy that included the use of horses. This particular ranch had experience with combat veterans, PTSD, traumatic experiences, and many other areas of healing that military and military families alike could benefit from. Three months of horse therapy typically equated to three years of traditional talk therapy. Skeptical as I was, something drew me to this place. As we drove through the property, I could feel that it was a place of healing.

I could see the pasture with about five horses all gathered in the center as they enjoyed their afternoon lunch of hay. As we entered through the gate, the massive red barn and covered arena were to our right. The ranch was obviously very well maintained and we were eager to see what this was all about.

Our first sessions were more of a meet and greet for Gentry, Kylee, and I; each with our own horse therapist. "So, what exactly do we do?" I asked.

"Well, we'll all go out into the pasture and we'll let the horses pick you. See, when you go out there, you'll think you are picking the horse, but it doesn't work that way. The horse must pick you," Callie, my assigned therapist explained. *I don't know if I'm ready for this.*

Gentry and Kylee were ready to get out there and pick…ahem…see which horse picked them. I was fearful and decided to wait a few sessions before I interacted with the horses inside the fence.

I vaguely opened up about my experiences as a child, my upbringing, my brother, and his death. I wasn't really expecting much, as this seemed quite unusual for me to be talking about my life next to a fence with these massive, beautiful animals inquiring as to who this new two-legged stranger was amongst them.

After a few sessions, I could see that both Gentry and Kylee were eager to get to horse therapy each week, both with different experiences that must have really challenged them. I, on the other hand, was not so eager. Callie encouraged me. "Renee, it's time for us to go into the pasture. Let's see what happens."

"Alright. Let's see," I replied reluctantly.

"When we go inside the fence, I want you to just look around, walk slowly and take in your surroundings. I want you to be aware of what you are feeling. You see, the horses can feel human emotion. They are mirrors of what you are feeling. One of the huge benefits to this type of therapy is that you can see how you are feeling through the eyes of your horse, even when you may not be able to distinguish what you are feeling on the inside. It brings awareness to your feelings so you can begin to change and recognize your emotions."

This all made complete sense to me and it was quite fascinating, but I was kind of a "I'll believe it when I see it," kind of girl. Skeptical, perhaps, with a side dish of some cynicism. We entered the gate together and I saw four of the horses prop their heads up and take note of me.

"So, which one of these horses do you think you want to attempt to be introduced to?"

I saw the light-colored male horse take interest in me and he began to approach. This beautiful horse had a golden mane that bounced in the sunlight as he walked towards me. I immediately took a liking to Tristan. I felt comfortable with him and I could see that he was quite friendly.

I looked around and saw a huge black male horse in the back, keenly aware of our presence.

"That's Rex," Callie stated. "He's the boss in here. You need to be careful around him especially with other horses around. He likes to run them off."

I had no plan to get too close to Rex. His size alone terrified me. In the light, he appeared midnight in color and his height was intimidating. I focused back to Tristan and we got to know each other a bit, made a little small talk about my past and it seemed that very quickly our session was up. *Ok, I can start to see the appeal.* I felt comforted and safe being able to share with this soothing, beauty of a gentle giant by my side.

The following week I was quite eager to get back to see Tristan. I felt we bonded well and I don't bond well with others very easily, so this was progress. My guard was up and for good reason. "Renee, let's head inside the fence. I have a challenge for you today."

Great. I smiled. Callie continued, "See Rex over there? I want you to take this harness and walk over to Rex and harness him."

"You want me to do what?" I exclaimed. "No way."

"What's wrong with Rex?" Callie asked.

"Well, he's massive and he's scary looking, like a black stallion that is mysterious and intimidating at the same time. That horse could trample me."

"But, he won't," she said.

I began to slowly approach Rex. My heart was beating so fast, I could feel it outside my chest. *This is crazy. What am I doing?* I got about three feet from Rex and he walked in the other direction away from me. *How the hell am I going to get this horse?* I approached him again and he walked off again.

Callie asked, "What's happening here?"

"Well, this horse obviously doesn't want to be around me," I said.

She replied, "Well, maybe he is just showing you how you feel about him."

Ah ha! I approached him a third time, apprehensive and afraid. I was about to touch his face, and he walked away from me again. I stood there, abandoned. I was abandoned by a horse that was supposed to be helping me and was doing nothing but frustrating me. My assignment was that I MUST harness this animal.

I stood there by myself and started to cry. The horse who is supposed to be helping me, didn't like me. As a matter of fact, he completely left me. Every time I approached him, he left, completely disinterested.

Callie asked me, "What are you feeling right now? Be honest."

I stood there, trying to put my feelings about this stubborn horse into words. "Well, I feel rejected by him. I feel anxious that he will hurt me, and I feel like I won't be able to accomplish the task at hand. I am completely worthless to this horse, not only that but I feel like a failure."

"Ok," she said. "Now go get your horse."

I stood there and stared at Rex. My hands trembled with both shame and disappointment in myself as I began to approach Rex. In that moment, I was broken. I was real. I was honest. For the first time in a long time, I was able to acknowledge my brokenness. Rex turned towards me. He approached me and placed his head on my shoulder. I leaned into him. I slowly raised the rope harness and place it around his head and tied it. I was now able to grab Rex by the reins. *I did it!* I was elated. I couldn't believe I just did that. I felt as if I had just conquered Mount Everest.

"Ok, now I want you to walk with him as a partner. You are partners in this journey and he needs to walk beside you." Callie said. I started walking and Rex began to pull me. Rex was walking me. Obviously, I only thought I was boss for a moment, and then reality hit. I realized he was the boss at that moment.

"Gently pull him back so he is walking beside you. If he doesn't listen, pull a bit harder."

I was able to get Rex to walk beside me with tight reins, but…he was beside me.

During each session, I grew more and more confident as I approached Rex. He felt it and I felt it. Sometimes, getting started was the hardest part. I had to face my fears and when I did, I realized Rex was not there to hurt me. He was simply teaching me.

Session after session, I grew in confidence. Rex and I became good friends. He knew my deepest and darkest pain and he loved me more for it. He helped heal my marriage and my relationship with my daughter. He taught me things about myself I could never face alone. Rex, the big, scary, horse I was terrified of, saved my life… and I am forever grateful.

It was time for Rex and me to part. I knew it, and he knew it. As he slowly kept up with my pace, walking beside me without even having to ask, we both knew this partnership had come to an end. It was time for me to take the lessons I learned here and move forward in my life. And it was time for him to make a new friend and help them.

2017 - The horses of PAWS for Reflection equine therapy,
curious as to why one of them gets to be outside the fence.
Courtesy of Renee Nickell.

Things Will Get Worse Before They Get Better

Though I learned many new things during horse therapy, I also knew it would still take time to shift the changes into our home. Callie had warned me, "Things will get worse before they get better." What she meant was when you begin to implement change in your home, there will be resistance because it is different from what the family has been doing for years. We were retraining ourselves how to be better parents, better communicators, better friends, etc. There was fear in moving forward…in a different direction than where we were previously headed. We had been headed for destruction, but God had a different plan for us.

I thought things would be smooth sailing, things would be perfect. But our reality was far from that expectation. We were dealing with a seventeen-year-old girl, hormonal, unsure of herself or her future, and struggling with PTSD. While horse therapy did wonders for my self-confidence in teaching me how to express myself, I lacked the ability at times to overcome the pain from past situations. Especially those I couldn't fix instantly, and the pain of losing my brother.

Because we finally began to deal with all our suppressed emotions, it seemed the fighting got more frequent and more intense. Gentry and I were finally a united front, and Kylee knew that. She did what any hurting teenager would do and she rebelled against us stronger than before with all those old emotions coming to the surface. Now, my definition of her rebellion was much different than my own actual experience as a teenager. When I was seventeen, I was already drinking and smoking cigarettes. I had the most horrible "friends", who weren't really friends, but accepted me. In hindsight, they were only trying to find their way too, but not beneficial to me and my circumstances. Kylee was a good kid, but just lacked the emotional stability to handle stressful situations, and she revolted against us… with her mouth.

I didn't realize how difficult it would be for all of us to finally face the pain in our lives. I didn't really believe Callie when she had told us things would get worse before they got better. I didn't want to. I couldn't understand why we had to keep walking through all this pain when we had already withstood so much.

I think it was too hard for me to face the horrible fact that I couldn't help my children the way they needed me to. I came to a breaking point. After the months and months of horse therapy, I broke. Gentry asked me what was wrong. It must have been so obvious by the expression on my face. "Nothing is wrong with me, I just can't ever really 'fix' anything," I said.

"What do you need to fix?" Gentry inquired. I didn't really know. I just felt like everything was broken and falling apart; unraveling like a spool of thread. I pictured myself like this: As the spool dropped out of my hand, I quickly tried to grab it before it completely unraveled. But the more I grabbed the string, hand over hand, the more (and faster) the thread completely unraveled. Trying to wind it back onto the spool felt nearly impossible. And even if I did, it wouldn't look anything like that perfect spool I once held in my hand. I felt my life was the spool of messy, knotted thread.

I could feel the pressure inside me. I didn't want to talk, but I knew I would burst at any moment if pressed. Gentry, who always wanted to fix every situation, or at least get me to talk, began to press. I couldn't handle any more pressure, and I exploded on him.

"What do you want from me? What do you expect from me?" I yelled.

"I just want you to tell me what's going on with you!"

"What do you think is going on? Why do I always have to say it?" I replied.

"Renee, I don't understand."

"This week…this is the last week we spent with Sam. I can't even cope. When will this end?" I told him.

"You don't have to push me away! Why don't you just tell me?" he asked.

"Why don't I tell you? Why should I have to?" I questioned angrily. I could feel an argument stirring.

Why did I need to tell him that there will always be days? There will always be Memorial Day, his birthday, our last week together, the last call, the last hug, the last voicemail, Veteran's Day, Thanksgiving, the anniversary of his death, Christmas. Why do I have to remind others just to validate my ineptness to "be happy" during those times?

"Renee, you always do this. You think I'm just supposed to know how you are feeling. You always push me away," he said firmly.

"What do you want me to do?" I yelled. "Do you think I want to always talk about how hard it is?" I replied.

"You don't have to do this. You don't always have to act like I'm the enemy. For Pete's sake, Renee, I'm not your father! I'm not going to leave you! Stop treating me like I am!" he yelled.

I stood there, dumbfounded. He was right. I hated any form of confrontation. I had been taught my whole life to not share my feelings. My father never wanted to face that our family wasn't perfect. Now I had a husband who wanted to hear me and wanted to help make things right. It was hard enough in horse therapy, but now I had to put it into practice, into real world situations. The hard truth was that I feared losing those I loved...even my husband, for simply expressing my feelings.

How am I going to get where I need to be? I knew I couldn't deal with this alone. I'd been trying to make my own way out of fear. I knew I needed to seek God. It was that pivotal moment when I realized I could keep suppressing the pain, or I could ask God for help. I needed to find purpose for my life. I spent a lifetime feeling insignificant, like I was always runner up and never a priority. When I was in my youth, I made myself a priority through rebellion with the aim to seek the attention I needed until even that didn't do the trick. I connected the pattern. I had always needed my brother's footsteps to follow, and now I had nothing to make me feel worthy or significant. I was lost without him. I was afraid of losing. I was afraid of failing. I had given up inside, and now it was time to find my "why".

Where is God?

I imagine this is a question that is asked more often than not when a person grieves the death of someone so loved, someone so kind and loving. You wonder why they had to be taken. There were many questions I asked myself over the years after Sam was killed. There were times when I felt so alone, I felt like God didn't exist at all. Thankfully, I had enough history with Him in my life that reminded me that He did and does, in fact, exist.

Regardless of what I *felt,* I knew God was still God, still the same, never changing, ever present and merciful. There are many things I didn't understand after Sam's death. Please have mercy on me, as there are those that think we didn't lose Sam because we know he is waiting in heaven. Clichés aside, I lost my brother due to death. I will not be with him until eternity. This does not discredit his existence in heaven. As a Christian, there are plenty more clichés people say to make other's feel better…or themselves. I'm unapologetic for being a bit cynical, as the pain of losing my brother is real and my faith in God struggled greatly. I wish I could say I was one of those whose faith never wavered, but I can't. My faith did waver.

I could not fathom why God would take someone like Sam from this earth. He was a man who loved everyone and everyone loved him. He was the guy who would go to the dining hall at Penn State and make the tallest ice cream cone he could, and randomly take it to someone he didn't know. Who does that? Well…my brother did.

What I have had to learn through the pain and the heartache, through the trauma and the hard days…the days I wasn't sure I would make it, is that life didn't end the day Sam died. If it had, I'd be with him right now and so would everyone else who loved him. That's not how it works. We are left here to grieve, to work through the pain, so God can do a work in us and in others that would not happen otherwise.

I had to learn to live on purpose and with purpose…a kind of purpose I never had before. I was a wife, I was a mother, I was a daughter, but I did not know my purpose. Isn't that what we all are searching for? Obviously, I am not the only one. The book *A Purpose Driven Life* by Rick Warren has sold over 34 million copies. After Sam died, I questioned everything. I needed to discover what God wanted from me and what I could take away from all the pain so that Sam's death didn't make me bitter and angry. I had already spent a lifetime in bitterness and anger. I felt rejected by my father and I felt like I could never live up to anyone's expectations of me. I had to willingly, and intentionally, put myself back in the game.

So, I learned to live differently, differently than I had ever lived before. I had to discover who God the Father was to me and allow Him to walk with me through the healing process. I began reading book after book on forgiveness. I became a sponge. While my family was sleeping soundly in their beds, I would stay up hour after hour reading about what it meant to be a vessel of forgiveness towards others and towards myself.

I learned a lot about what forgiveness is, and even more so, I learned what forgiveness is not. You see, I lost who I was without Sam. I was never just Renee and I didn't know how to navigate my life or my identity without him. You don't just go on the same way you did before. You must learn a whole new way of living and that's what I had to do.

There is something beautiful to be said about redemption. There is still redemption in the loss, regardless of whether we see it, when we see it, or if we even choose to see it. I remember praying to God after Sam's funeral that somehow, I could make a difference. I didn't know how. I didn't want to make a difference for my own benefit, but I wanted to know that somehow, someone's life was changed because of Sam's death. I prayed that someone would be affected in such a way that the trajectory of their life was changed. It was my heart's cry for a good two years. I received a phone call out of the blue.

I had a toddler on my hip, while helping my daughter Leah do schoolwork when the phone rang. "Hello?" I said.

"Is this Renee, Sam Griffith's sister?" I was asked.

"Yes, who is this?" I replied.

"My name is Jaime. I was a friend of Sam's. I live out in Phoenix." Jaime paused. "I know this may sound odd, but I had to call you and tell you something."

"Ohh-kaay," I listened, slightly confused.

"Well, I was at Sam's memorial service in Virginia. Unfortunately, you and I didn't get the opportunity to meet in person. I had to tell you...your words...well...the eulogy you gave had a tremendous impact on my life. I changed my walk with God. I had to call and thank you for your inspiring words on what I imagine was one of the worst days of your life. I wanted you to know that God gave you incredible strength that day for a reason. I am certain I am not the only one who was inspired. I hope to one day get the chance to meet you in person."

"Jaime, I don't know what to say." I was stunned, on the verge of tears. I had doubted everything in my life, I had felt hopeless and meaningless. Then unexpectedly, this person who doesn't even know me, tells me that God used me to help change his life. A person I had never met before.

Jaime and our family instantly became fast friends. We are still friends today, and I don't mean just social media friends. I mean, he flies to Dallas occasionally and we always meet up for dinner or catch a Texas Rangers game.

Not long after that chance phone call, I sat on the floor of my bedroom closet and decided to pull out pictures from way back when. They were from a time Gentry and I had lived in Phoenix, about 10 years prior to ever knowing who Jaime was. I came across some photos. Sam had come to stay with us and he took Gentry out for some beers and met up with a friend…Jaime. Sam, Gentry, and Jaime had hung out all night sharing stories and beers. I couldn't believe my eyes. I probably wouldn't have believed it if I didn't have the pictures to prove it.

2003 - Jaime, Gentry, and Sam during a night out in Phoenix, AZ.
(I didn't meet Jaime until years later, after Sam's death.)
Courtesy of Renee Nickell.

I don't always understand the things of God. I have tried long and hard to figure Him out, and I can't. I don't understand why bad things happen to good people and I don't know why families break up, but I do know that if you allow yourself to be changed, God can and will use you in ways you never imagined for yourself or for your life.

Sam's death has changed the way I live my life. He has changed how I view friendships. I love more and more deeply. I stop to smell the roses, as cliché as that is, I do. I stop and look at my children's drawings even if it's the tenth one they brought me that day. I always tell them they are amazing...because they are. When I look at the stars, I look for the twinkling ones that shine a little brighter.

Sam's death could have filled my heart with hate and bitterness. I had every reason to become a hateful, vindictive, resentful, unforgiving person, and you know from reading my story, I was headed down that road. Trust me, I did not wave a magic wand over my head. I had to make a decision. God is so sovereign, He allowed me to make that choice. I knew Sam wouldn't ever want his death to fill my heart with hate; therefore, to honor not only Sam, but God, I had to let God chisel away the pain.

I learned to trust God again. I'm not saying I'm perfect because I still struggle. There will always be bad days where I want to throw the covers over my head. I may always struggle with why I was never good enough for my father's love...but those days are less and less. While I may have been a disappointment to my earthly father, I found another Father who saw the ugliest parts of me and continued to love me through the process. I sit here and smile...because Sam knew the ugliest parts of those closest to him, and still chose love.

There is no way Sam would have ever disowned a single soul. It wasn't in his nature. I swear it seems supernatural...his ability to love people regardless of their actions and mistakes. I want that ability. So, I daily try to choose better, to give more grace and mercy because God knows I usually need a double dose of it myself. Aren't we all in need of second chances, or third, or fourth? We just keep trying to get it right until we do. That's the beauty in failure. It's the not giving up, it's the persevering after falling flat on your face time and time again that changes our hearts and character for the better.

Kylee: A Pillar

Learning to give more grace has not been easy, especially as a mother to a teenager. Saying I never wanted to be like some influential people in my life and actually not being like them are two completely different things. There are attributes that are instilled in us as we grow up whether we like them or not. Sam and I both had to take the good with the bad...and we both had to work on it. He wasn't perfect, nor am I.

One of the greatest desires of my heart was to raise my children in a calm, nurturing, happy home. The picture I had in my head wasn't exactly my reality. I took my hurts into every area of my life and I struggled greatly as a parent, trying to make the best decisions for my children. I know Kylee hates when I say this, but she was our practice child, our firstborn. We made the most mistakes with her, but we also learned to make the biggest changes in our own lives because of her.

As stubborn as she has been (since her birth, I say this in love), Kylee is destined to change the world. Kylee has been a pillar more times than I'd like to admit. She experienced tremendous trauma the day her uncle was killed and it changed who she was. She watched as her mother lie on the floor screaming in shock, she shushed her little siblings in comfort and hid them away to protect them. She called her father and so many others to calmly tell them the news that Sam was gone…she was twelve.

The years following, she was in just as much fog as I was, only I couldn't see past my own pain. She rebelled as any child would do, unable to cope with PTSD when I just wanted our family to be "normal". It's been a long, broken road and I have had to learn to love her through her mess, just like she has me. At times, we took drastic measures to "help" her and it wasn't what she needed. She wasn't rebelling to rebel, she was hurting and didn't know how to express it.

As if teenagers don't have enough to deal with, add trauma onto that, as well as her chronic illness she has dealt with since she was seven. It was a recipe for disaster. It was a disaster. Our lives were a disaster for years following Sam's death. Sometimes all we could do was just float and let the waves carry us as they splashed our faces. As soon as we started to get tired and weary, or felt like we would succumb to the waves of the ocean, we found a little more strength to keep our heads afloat. That was how our family functioned. We grasped at any possible means to make things better.

How do you make things better when you don't really know what better is? I had no idea what a normal, functioning family looked like. I certainly had no experience as a child. Maybe families live in dysfunction. It's *their* normal. I wasn't satisfied with my idea of normal. So, I began to clean up my life. I had the tools to make things better. I was the parent and I had to make things better with Kylee, even if she made wrong decisions. For so long, Kylee and my roles were reversed because I had been absent. It was time for me to put my big girl pants on and be there for her.

This is where unconditional love was challenged. We said we had unconditional love for our children, but then we placed stipulations on them. Though it wasn't necessarily spoken, we expected that they must perform well, obey, get good grades, be respectful, go to the college of our choice, and have great friends... *THEN* the environment in the home would be a wonderful, magical place. No. Uh uh. I learned it doesn't work that way. I had it backwards. After the hours and hours of therapy, the years of bad advice from well-meaning professionals, the gazillion ways of failed discipline tactics, the horse therapy, the yelling and screaming... I realized my love had become conditional. "I will love you if..." A parent/child relationship will never work with conditions when it comes to earning love.

I had to relearn how to love my daughter in a way that she felt safe and secure, and I had to be strong enough to place boundaries and consequences on her. I had to not only let go of the grudge, but say, "Hey, let's hang out this afternoon."

Kylee made a lot of bad decisions following the death of her uncle. So did I. I didn't want the world to judge her any more than I wanted them to judge me. And so, our circle got smaller and smaller. I no longer had this desire to please everyone around me, or to be this person who everyone had to like in order for me to be happy. I learned you can love people right where they are. Those that are hurtful and mean, you can pray for, walk around with caution... and love from afar.

Life certainly was not perfect for us, but I knew without a doubt, it could be happy. We took the good with the bad. Not every day would be a bad day. We dealt with each situation right in front of us and let the rest of it just fall away.

Kylee taught me a lot about being a mother. I was a young 22-year-old holding my baby, saying "I will never...". It didn't work out so well, but God was so gentle with me. He allowed me to become the person, the mother that I always dreamed of. Do I still compare myself to others? Yes, sometimes I do. But when I see that well put together Mom, who is fit and dressed impeccably with her well-mannered children in Walmart when suddenly, her two-year-old begins screaming and throwing herself on the floor…I feel relief. *Ok, it's not just me. I am still doing a good job. I am a good mother.*

I am fierce and protective. If anyone hurts my child, I am cut to the core and my gut reaction is to want instantaneous justice. My mom taught me from the time I was a little girl to fight my own battles. She made Sam and I stronger because of that. Sam just learned faster than I did. Fighting my own battles usually erupted in an all-out war. From a very young age, I learned to function in protection mode. I had a very good reason to be that way, but it didn't make for healthy and happy relationships, especially as a parent.

Kylee and I had to find peace as mother and daughter. She had to know that regardless of what she walked through, I would be there for her. I think we have found that peace. We still get frustrated with each other, but now, we get through our crazy quarrels. I tell her what I think she *should* do, and she tells me what she *wants* to do. Sometimes we compromise and sometimes we don't. Sometimes I win and sometimes she wins (when I realize I need to let go of control).

Regardless, she's my daughter and I would fight for her no matter what. That's what parents do. That's what parents *should* do. Regardless of their choices, unconditional love means just that, unconditional. It's not enabling but loving regardless of what I think is right. Good Lord, we'd all be sent straight to hell if we didn't have a God who loved us unconditionally, especially when we make mistakes. *Thank you, God, for the cross of Jesus!*

Loving our children may be forcing them to seek the help they need to get better after a bout with depression. It may be interfering and being nosey. A parent must know who their children are spending their time with. It may be telling a boy to get lost, that he's not good enough for her, and to come back later after he gets his life together. It may be accepting that your family isn't perfect, that it's a little broken, but repairable if you're willing to accept the challenges and face them with truth.

The one thing I want each of my children to know is this: Their father and I, no matter what challenges they face (and we've been through A LOT), will never leave them, abandon them, or love them any less even if we disagree with the decisions they make. It has been hard, but when the children are upset with us, or perhaps our reaction, they have learned to tell us. When they do, we listen and try to do better next time. Parents need to hear their children. Yes, you are the authority, but if your words or actions are hurting your children, change your words or actions. God doesn't smite us every time we mess up.

Kylee. She just makes me a better me. I'm pretty darn sure Sam would be proud of who I am. I'm pretty sure he always was.

CHAPTER 13
MOVING FORWARD

The Wreath

I went into my closet and pulled down my brother's flight bag. Inside was his desert camouflage uniform. I unzipped the bag and slowly pulled out his uniform. I raised the shirt in front of me. His name tape "Griffith" stared me in the face. I placed it against my chest and cried. I previously had a plan to use his uniform and make a wreath, but I could never allow myself to cut it. I realized…it was time.

What was I going to do? I was going to make two wreaths and give one to my mom for Christmas and one to my estranged father… well, that was the plan. I held that shirt and stared at it. I couldn't cut his shirt. Perhaps I could just cut his pants. I went online and ordered a few "Griffith" name tapes to attach to the wreath after it was completed.

After gathering the supplies, I slowly sat down on the living room floor, Sam's pants in my lap. I held a pair of fabric sheers in my hand. The hardest part would be the first cut and I knew that. My children were standing in the background watching, almost as if they were holding their breath with me. I looked over my shoulder and tears had filled their eyes, as they just silently observed. They knew this was hard.

My hands were shaking and I fought to keep my composure. I placed the sharp scissors against the fabric, took a deep breath, and made the first cut. A tear dropped onto Sam's pants. *I did it.* I imagined Sam standing there smirking at me, probably asking me what the big deal was, it was only a pair of government issued pants. He'd probably say, "Cut the hell out of those things!"

So, I did. I shredded those pants into one-inch strips. I tied every single piece to the wreath wire, my hands in physical pain from the inflammation from my autoimmune disease. This was more than just making a gift. This was an act of love. I knew my mother would understand this. I hoped my father would too.

By the end of the week, I had two beautiful wreaths made with red, white, and blue fabric, and Sam's uniform. My fingers cramped in pain, but my heart was full at such a major accomplishment. I knew I was stronger than I thought I could be.

My mom would be flying in soon. We always made sure we were together every year on the anniversary of Sam's death. December 14th was approaching and I wanted to make this day special. Typically, we always went out to dinner. We'd have nachos and chicken wings and drink a Jim Beam and a diet Coke. We would toast to Sam and the legacy he left to so many. We remembered his sacrifice and we remembered his love. Laughter and tears abounded as we remembered his life and what he meant and still means to all of us.

Mom was sitting at the bar in our kitchen after a long, already emotional day. The kids were all standing around her as I walked into the kitchen holding her wreath. I walked up to her and presented it to her. Stunned, she had no idea what was happening. She looked at it and began to weep, holding it close to her, softly rubbing her hand over Sam's uniform. Everything I had experienced over the past week was worth the joy I brought my mother, as hard as it was. It was another step in letting go and I knew I was headed in the right direction.

I placed the other wreath I had made in a wreath box, unsealed. I didn't know what I would do with it. At that point, I hadn't seen or heard from my father in four years. What on earth would I say? What could I say? As I sat and pondered, I realized I wasn't ready. I was still holding on to past hurts. Part of me wanted to send it and part of me didn't think he deserved it.

I decided not to send it. Not out of spite. I was right, I just wasn't ready. As I sat and stared at that box, I wondered if I'd ever be ready to mail that to my dad. After all, I was giving away a part of Sam's uniform to a person I knew cared nothing about me. I took the box and placed it under the stairs, in the back corner of the closet. And there it sat…waiting.

2016 - Mom with the wreath I made from Sam's
uniform for Christmas; a true labor of love.
Courtesy of Renee Nickell.

I.C.U.

It's crazy how sometimes something can happen to completely shift your focus. I believe sometimes when we are on the cusp of breakthrough in our lives, teetering between holding on to those things we should let go of and being free, God allows something to happen to completely shift our thinking. For me it happened at about 2 a.m. one early Monday morning, just two months after Christmas.

My eyes flew open out of deep sleep on that cold February morning. *What is that feeling in my chest?* Panic came over me. I'd had a few panic attacks in my life, but nothing like this. My heart was racing. We owned a pulse oximeter since Gentry's heart surgery a few years prior. I sat up in bed, reached in my nightstand drawer and pulled out the pulse-ox and placed it on my finger. My heart rate was 160 bpm, and then it'd drop to about 50 just one beat later. I thought I was dying.

"Gentry, wake up! Something is wrong with my heart!" I said in a panic.

"What's happening?" he said.

"I don't know. My heart is bouncing all over the place. I need to go to the ER." I sat there a moment.

"Wait, I'll have Kylee take me. You stay here with the little ones in case they wake up. I don't want Maggie to be afraid with us both gone."

Gentry in a sleepy fog agreed. "I'm sure it's nothing. You'll be fine. Just try to relax."

I awakened Kylee and she quickly drove me to the freestanding ER down the street. As soon as I got there they hooked me up to monitors. The doctor said, "You are definitely in atrial fibrillation. Do you know what that is?"

I replied, "Yes, my husband has it."

He gave me an aspirin. "Well, we need to transport you to a major hospital. You'll have to be admitted."

I got scared. My heart was racing out of control, bouncing back in forth from low to high. The chambers of my heart were beating at two different rhythms. I looked at Kylee. She had tears in her eyes.

"Kylee, don't be scared. I'll be fine."

"I know, Mom," she replied.

We both sat there in silence for several minutes. I'd never experienced anything like this. As I sat there, I wondered if this was how I'd go, never really making a difference in life, never mattering. I had so many unfulfilled plans for my life, for my kids, for our family. I looked over at Kylee…"Kylee, I'm really sorry for all the times I could have been a better mom," I started.

"Mom, seriously. It's ok. Just stop."

"No, I need to say this. You've been so strong for me. After Sam died, you went through so much and I wasn't there for you. I just want you to know I'm sorry, and…I see you."

Kylee eyes welled up with tears again, and she walked over to me and gave me a hug. "Mom, you're going to be okay."

The paramedics walked in with a stretcher to transport me. They loaded me up and we began our route to the hospital. The paramedic who was in the back with me needed a swift course on bedside manner. "You ever have this before?" he asked.

"No, this is my first time." I didn't even want to talk. I couldn't focus on anything except my heart beating out of my chest.

"Well, they'll probably just shock your heart and you'll come out of it."

Wait, what?

He continued, "My coworker used to go into SVT (supraventricular tachycardia) all the time. Yeah, her heart would beat over 200 bpm."

This is NOT helping. Would this guy just shut up?

"She was a pretty tough girl. She'd actually just shock herself. She'd have someone there and she'd put the paddles on her chest and shock herself."

I literally could feel my heartrate increasing.

"Don't you worry, they usually put you to sleep for it."

I now officially feared for my life. They were going to shock my heart, to stop it and restart it. *This is it, I'm a goner.* I got settled in a room. Kylee went home. They immediately placed me in ICU. Blood tests were ordered and all types of heart tests. I just lay there alone. Being in A-fib literally feels like you are running a marathon except when you are racing, you can stop to catch your breath. My body thought it was running with no break to rest. I was out of breath and so tired. It was like trying to take a nap in the middle of a race.

I just laid there and pondered my life. I thought about Sam. I thought about all the decisions I made in my life…the good and the bad. I prayed. I ask God to please let me have another chance to make things right. I thought of my dad and how much unforgiveness I still held in my heart toward him; not for anything in particular he had done, but perhaps what he could have done and didn't.

I texted my cousin Karen. I knew I could be transparent with her and she'd understand:

> "Karen, I'm in the E.R. My heart is racing. Doc told me if I don't come out of it, they'll take me to the O.R. and shock my heart. I just wanted to tell you…Look, if I don't make it, will you just tell my dad I love him and I'm truly sorry I was always such a disappointment to him. Just please… tell him that."

> She replied, "Renee, I'll be praying for you…and no, I will not tell your dad that. I'm believing you're going to be fine and I also believe one day you'll be able to tell your dad that you love him yourself."

About that time, Gentry walked in. He came right over to me. I put my face into his chest and cried. "Renee, you're going to be ok. I promise."

"Gentry, I was never good enough. I was never good enough for my father. I wasn't always a good mother or a good wife. I've failed so many times. What have I accomplished?"

"Renee, stop believing those lies. We all make mistakes. You've been a wonderful mom and wife…and daughter. How about I call your mom and ask her!? I know what she'll say," Gentry encouraged.

"But…maybe God made a mistake…maybe it should have been me. Sam was the one who always did great things. He was the one everyone loved. He should be the one who is here," I said.

"God has you here for a purpose…a great purpose. I don't know why God chose to take Sam, but I do know He doesn't make mistakes. You have a lot of people who love you, and those that don't? Well, they are just missing out. Maybe God wants you to see that there are those He has placed in your life that do love you," Gentry said.

"Maybe so, sometimes…I just wish he was here…I just wish Sam was here so I could talk to him. I just miss him so much. Will the pain ever go away? What if they have to shock my heart?" I said.

"They won't." Gentry prayed, "Father, we plead mercy over Renee right now. We ask You to convert her heart. We ask for no further issues with her heart. Give her peace. In Jesus Name. Amen."

I just laid there and tried to rest, thinking of my children at home, asleep in their beds. Nine hours passed and my heart was still racing.

The doctor came in. "Mrs. Nickell, we're getting the O.R. prepped to take you down. We'll sedate you and then we'll convert your heart for you."

I was terrified. The doctor left. I sat up, placed my legs over the side of the bed to get out and walk to the restroom. I leaned forward just a bit and something happened. I sat there a minute. The monitor that was keeping track of my heart that had the words "A-fib" on the screen, now showed normal sinus rhythm. I said to Gentry, "I've converted. Look at the screen. I spontaneously converted."

The nurse came in, stood there and looked at the screen with her finger up against her mouth. She seemed a bit shocked. She confirmed that I had converted. I felt the Lord tell me that it was HE who converted my heart. He restarted it, in a new rhythm. It was as if He released me from so much unforgiveness. I went home. Later that week I discovered I was severely anemic and on the wrong medication which caused my heart to go into A-fib. After several IV iron infusions, I was feeling as good as new…and it was time to write some letters.

Letting Go

Months passed. Spring was in full bloom and it was time to do some closet cleaning. I felt I had a whole new outlook on life. I don't know what happened in that hospital except I had more than just a heart conversion. I felt compelled to forgive people who had hurt me and who I knew felt hurt by me too, justified or not. It's never right to retaliate, even when you feel you are justified. I'm sure they felt just as justified in all the fighting after Sam died, just as much as I had.

I opened that closet door and began moving things out. Cleaning out closets is always therapeutic for me. It's a way of purging the things of the past that can weigh us down, a means to make room for the new. Old jackets got thrown in a pile. Those bins of children's clothes got pulled out to store somewhere else. As I reached to the back of the closet, I saw it. There was the wreath I had made for my dad but never sent. It was staring me right in the face. I had forgotten about it. As I stood there holding the wreath box, I knew it was time to do more than just clean out closets. I needed to dust off the cobwebs of the dark places in my heart and extend an undeserving act of love. I did, in fact, love my dad, maybe too much. Perhaps I had actually given him too much grace over the course of my life and set myself up for heartache over and over. I felt this wasn't one of those times. I needed to do this.

I wrote four letters to people I believed I had hurt and who I felt had hurt me. I emailed two of them to the only email addresses I had from years prior and mailed the other two. I knew I had to just let go. I had to let these people who had hurt me know that I not only felt forgiveness towards them, but that I was sorry for what I had done to hurt them as well. I don't mean the, "I'm sorry, but I'm still going to be vengeful, merciless, vindictive, and slanderous towards you." I felt it in my heart. I wished them well. I wanted God to bless them and help them find peace in their lives. I wanted that for them, and still do.

I never expected responses and I never got them. Perhaps they never even read the letters. I was okay with that. I didn't need a reply. I knew that just the act released me from years of pain, resentment, and anger. I knew it wasn't my position to vindicate myself. Regardless of what others believed about me, I knew I did the right thing and my goal now wasn't reconciliation, it was freedom. Those two things are completely different. I had them confused for so long. I felt I couldn't be free unless they apologized and we reconciled. God showed me that wasn't the case at all. True forgiveness happens when you can be at peace with just...forgiving.

And so, I mailed that wreath to my dad. I affixed the postage, prayed a blessing over it, and released it...finally. I don't know if he ever got it. I don't know if he ever opened it. He could have burned it, but I hoped he would just accept it because it had a piece of Sam attached to it.

There were many times as a Christian that I wished I could experience freedom. Freedom from resentment and pain. I had previously thought I could never get to that point. But, I finally kicked myself in the ass, stopped feeling sorry for myself, picked myself up, and said, "I'm ready, God. What do you have for me?" So...ok. Well. Where do I go from here? How do I move forward? I waited my whole life for this moment.

I did not hear God open the floodgates of heaven and rain down his almighty powerful voice and tell me what to do. I wished it was that easy. I needed to figure out what my next steps were. I imagined when I finally felt free from my hurts from my past that I would, like Julie Andrews from The Sound of Music, run and twirl around in the mountainous meadows of Switzerland and sing at the top of my lungs. Um, it didn't quite happen that way.

I also thought, somehow, the pain of losing Sam would be gone. Nope, that was still there. That pesky thing called grief sneaks up on you at the most inconvenient times...still...to this day. Somehow, I miss my brother now more than ever. I knew he'd be proud of me, regardless of the outcome. He wasn't here to smooth things over like he'd always done in the past. This is the way it was and this is the way it had to be. You see, there aren't always happy endings in grief, in relationships, or in families.

Life proceeded to happen. Kylee was about to graduate from high school. "Mom, I want to invite Grandpa to my graduation," she told me.

"Kylee, I don't think that's a good idea," I replied.

Her face sank.

"I just don't want you to feel the same rejection I've felt. He's not going to come and I don't want you to feel let down," I continued.

"Mom, I still want to invite him…because…well because I would never want to regret *not* inviting him. I mean, I don't understand what happened and I don't have to, but I still love him. I still love Neil and Ben. I just don't want a life of regrets."

Gosh, she always amazes me. Sometimes I can't believe the stupid things teenagers do, and then she blows my mind with how kind she is towards others. "Kylee, you need to do what you feel is right. It may not be what I feel is right, but I can't always make decisions for you. I don't want you to make decisions about other people based on what I feel. That wouldn't be right," I stated.

"Mom, I'm going to send him an invite. You know that feeling you have when you go to school and find out that everyone was invited to a party except you, even if you didn't like the person throwing the party? Well, I wouldn't want Grandpa to feel that way regardless of whether or not he responds," Kylee said.

I got it. I did. And so, Kylee graduated that beautiful May Saturday, surrounded by her family. The only tears were from me, having to face letting her go. It was another step in the journey toward healing, for her and for me.

Trip to D.C.

Soon after Kylee graduated, I decided to take a trip. I needed to get away. I needed to visit Sam. I needed some direction. I called my mom and asked her to meet me in D.C. over Labor Day weekend to celebrate her birthday. She obliged along with my two favorite cousins, Pam and Karen and my Aunt Linda. We had a fabulous girl's weekend planned.

We had many profound experiences on our trip to D.C. I'd like to call them miracles, because there were some things I just couldn't explain. Miracles can be found all around us and since Sam died, I've had to look for the miraculous at every chance…and appreciate them. It's what gives me hope.

Traveling, albeit by plane, is quite exhausting. My mom and I got there a day before everyone else so we could have some alone time at Sam's grave. We knew we'd be tired when we arrived in Washington D.C. We can typically maneuver around the metro quite well. After all, we had spent a lot of time in D.C. over the course of many years. Our trips to D.C. aren't the touristy trips one thinks of when visiting our nation's capital. Our trips look much different.

We debated if we wanted to take the metro to Arlington National Cemetery (ANC) and walk from the metro to the visitor's center, which was about a half mile. We could then take the shuttle to Section 60 where my brother is laid to rest. Neither of us had ever "Uber-ed" but thought this would be a good time to start. Uber could pick us up from the hotel and drive us right up to Section 60, but we would have to consider the walk back which is about a half mile as well. Either way, we'd get our exercise.

Fortunately, we both had our ANC vehicle passes. When a "next of kin" is buried in ANC, they give you a vehicle pass to allow direct access. This pass allows you to bypass all three police checkpoints. They move closed coned off areas for you, then you drive into a separate entrance right past all the tourists headed towards the Tomb of the Unknown Soldier. Guards must stop tourists from blocking the crosswalk, allowing you to drive through right up to your loved one's grave.

Our Uber driver picked us up. I believe he was Haitian. I looked at my Uber app to see he had given nearly 5000 rides. He was a very nice gentleman, probably in his late 30's. We told him we were going to ANC. We said, "We have a pass. Can you please drive us in?"

He said, "Oh no. They won't let me in. You can't drive into the cemetery."

Again, we told him, "We have a pass, you can drive us in."

Slightly confused, he now approached the first check point. We handed him the pass to show to the first police officer. He reluctantly took it, still confused. Police immediately moved cones when they saw the pass. Uber driver said, "Wow! How do you get one of those?" as if we were some sort of dignitaries or something.

Mom and I looked at each other and both thought, "Buddy, you DO NOT want one of these." I think Mom may have said it out loud.

He continued, "I've never been to ANC. I've dropped people off, but in all my days, I've never been inside."

Mind-blown, right? *This man lives in D.C., has given over 5000 Uber rides, and has never been inside the gates of ANC?*

After the third checkpoint, he drove into the gates of Arlington and Mom told him to please drive to her son's gravesite and proceeded to direct him. He was speechless. His very first experience visiting inside the gates of ANC was to escort a Gold Star family to their hero's grave. It was about a half mile drive and I've never felt so much deafening silence. He drove so slowly. I could hear the slight gasp under his breath as he looked around. It was obvious he was astounded and humbled by what he saw. I could feel the awkwardness of the silence. He wanted to say something and maybe we were waiting for him to say something. I gathered that most likely this experience may have been his first and last time he would ever get to drive his own personal vehicle through those gates…and I hoped it was.

I imagine, as an immigrant, he had to have pondered the sacrifice made for him, maybe even appreciate his own journey a little more. I'll never forget that man and the reverence he had for us, and for all those who paid the ultimate sacrifice.

This was one miracle we witnessed that day. The Lord allowed, not only our pain, but also the sacrifice my brother made, to directly affect someone else in a way that wouldn't have been possible if it hadn't been orchestrated to do so. I will always wonder what that man thought. I will always wonder how that experience affected him, maybe changed him. No doubt he will always remember that experience, my brother, and our family.

There is so much division in our country, I can't help but ponder the idea that there are those that want their voices heard. They want others to see their hurt and pain and somehow understand something others haven't experienced for themselves. Then there are those, like myself, who are also hurting for other reasons. Feeling alone is a form of isolation because most do not understand or comprehend the sacrifice of losing a loved one in war. That day, there was nothing political, cultural, racial, or religious that separated us from that man. This man, a stranger, simply stepped into our world for a moment in time and he felt. We were simply...Americans. Isn't that a miracle?

2017 - Our trip to D.C. spending time at Sam's gravesite.
Courtesy of Pam Maggi Hockenberry.

Divine Intervention

Aunt Linda always loved me growing up and I was ecstatic she came along. To say my cousins and aunt are a recipe for a great time is an understatement. Sam and I cherished our time with them as children. My aunt was my godmother and I love her so, almost as much as my own mom...almost. After spending the day shopping and eating in Old Town, we decided we'd go to Arlington Cemetery the next day.

It was almost six years since Sam passed, and they hadn't been to his grave. I was glad we all would have that experience together.

We had two adjoining rooms, thank goodness, because we never stopped laughing. It was Sunday morning, our last day, and we needed to get moving…but first, coffee.

"EEEEEWWWWWW! OH MY GOSH!" screamed Pam.

We all came running. Now, let me say that we were staying in a very nice Marriot Hotel in Arlington. This was no roach motel. My mom went high class or she never went at all. We had stayed there several times in the past, but nothing prepared us for what we were about to see.

"You guys, look in the plastic coffee cup!"

We all hovered around as if we were viewing a science experiment gone wrong. As we looked down into the cup, it was the grossest thing we'd *almost* ever seen. There it was…it was like a train wreck… we couldn't look away. We couldn't believe our eyes. A brand-new coffee cup, still sealed with plastic, was filled with shavings. I don't mean pencil either. I mean…well, downstairs shavings. Apparently, someone felt the need to leave their personal bodily shavings from down under in our coffee cup, because it would be too decent to flush it down the toilet. No, let's just place it under another coffee cup and put it right back on the coffee tray. That was their great idea. I'd say the best part of waking up was definitely NOT Folgers that morning.

2017 - (left to right) Pam, Kathleen, Renee, Karen, and Aunt Linda, having dinner and laughs in Old Town, Alexandria, VA. *Courtesy of Pam Maggi Hockenberry.*

After we got management and housekeeping involved, this incident set us over an hour behind schedule to get to ANC. After we got the hair removed from the room, we took a walk down the street to pick up some wine and flowers for Sam's grave. Now, I wasn't a drinker, but I was with a bunch of ladies that were. They loved their wine just as much as I loved my Starbucks. Today was an occasion for some wine.

We called an Uber to take us to the Visitor Center and decided to take the shuttle provided for the families of those buried in ANC. It was just us and one other family. The driver had to go to Section 60 where Sam was buried and another section on the opposite side of the cemetery where the WWI vets are buried. He came to the stop sign and was about to turn left to Section 60, when he paused. He looked up at us. "Folks, I think I'm going to take you all to the back side of the cemetery first. Then I'll stop at Section 60." *Well, ok, then.*

He gave us a brilliant history lesson during our journey, dropped off the other family, and then proceeded to Section 60. As we approached, we saw several families walking around on this beautiful, late summer day. "Hey look! There's a blue Mustang. Wouldn't that be funny if that was Kelsey?" Pam said.

Kelsey was my sister and I hadn't spoken to her in years. My stomach sank. "I don't know what she drives, but I doubt that's her," I replied.

"No, I think that's her car. Isn't that her walking over there?" Karen said.

"Ahhhh, I don't think so. I mean, yes, that is where Sam's grave is, but I don't think that's her. I mean, what are the odds?" I said.

"Well, maybe God wanted us all here at the same time…I mean, what *ARE* the odds?" Pam said.

As I exited the shuttle I could feel my heart pounding. I looked at the license plate: Gold Star Family. *Oh my gosh*, I thought. *What if it is her? What will I say…after all these years?*

"That's Kelsey, I'm sure of it," Aunt Linda said. We were now about 50 feet away and the woman was crouched down in front of Sam's grave. As we got closer, I realized it was true. My mom and Karen held my hands, I could feel them shake. I was horrible about confrontation and was fearful of what Kelsey would say to me. I was done with fighting and wanted no part of an argument. We arrived at ANC at the exact moment my sister had.

She looked up. I couldn't imagine her surprise or her apprehension as she saw us all there together. I can't imagine how I would have felt. Waters were muddied a long time ago and we had once all been close…well, we were family. I probably shared the same expression on my face that Kelsey had on hers. I just stood there, not knowing what to do. My cousins started crying, hugging her, just so happy to see her after all this time. We made eye contact. I can't lie. It was awkward. I had thought about this moment a thousand times, wondering what I'd say if I ever saw her again. I had no words. I just stepped towards her and hugged her. She graciously hugged me back and left us to have some time with Sam. Although we didn't want her to go, I understood why she did.

We sat there at Sam's grave, drinking our wine, laughing. I was the only one that got drunk on a little single serve bottle. I cried my eyes out one moment and laughed the next. I loved these people. There we were, sitting at my brother's grave, together as we had been before. Only one thing had changed. Sam was gone. That was a hard pill to swallow.

Kelsey came back after about an hour of walking. We shared some laughs, invited her to dinner, and said goodbye. I somehow knew in that moment, this was another part of myself that was letting go. After all this time, I had found peace. I was okay leaving things the way they were. I was okay not arguing and not reconciling. There was just...peace. This isn't to say I would never want that reconciliation. I mean, she is my sister. After all the hurt between us, I felt peace and forgiveness and love. I felt peace because I only had love for her in that moment and God allowed me to take another step moving forward.

Sometimes, we can entertain what we would do or say to someone who we loved and had been separated from. I know I did. We can "say" we forgive someone, but we still hold on to the anger and hurt in our hearts. I didn't know if I had truly forgiven until I saw her face, standing there before me at our brother's grave. That's when I knew I truly had. I needed that confirmation.

2017 - Mom and Renee at Sam's gravesite, Section 60, Arlington National Cemetery.
Courtesy of Pam Maggi Hockenberry.

2017 - (left to right) Aunt Linda, Renee, Mom, Karen, and Pam at Sam's gravesite,
Section 60, Arlington National Cemetery.
Courtesy of Renee Nickell.

2017 - (left to right) Renee, Pam, Aunt Linda, Karen, and Mom toasting in memory of Sam.
Courtesy of Renee Nickell.

CHAPTER 14
CALLED TO WRITE

Blast from the Past

"When are you going to write that book?" I mentioned earlier that it was my friend Amanda that said those key words to me one day. Those eight little words struck a chord in me and I knew she was on to something. I knew I was ready. I knew I was on a journey that would lead me to a place of discovery, a discovery about things I never knew about Sam. The previous ten months of experiences, walking through the pain of my past and learning to let go, led me to the moment right here. I felt as if God himself had asked me that question.

The car was loaded down like we were about to do a world excursion, but really, we were only traveling 3600 miles over the course of two weeks. Some people may call me crazy to attempt this with four children, but it was important and necessary for me to take this journey for my book. Three hours passed and we were still in Texas.

That's one thing people know about the size of Texas. Once you start driving, you feel like you never leave our gigantic state. Our plan was to make it to Nashville, where we would meet up with Jason, our childhood friend.

The last time I saw Jason was three years ago in Washington D.C., when I bought two identical challenge coins. A challenge coin is traditionally a military coin the member presents as a way of identifying their unit. It helps to boost morale and military members collect the coins over the years. The unit coin is held typically in a wallet so when challenged, they have their coin to present.

The coins I had were remembrance coins for Sam. I had them engraved with his name, his call sign, and the most important inscription "BROTHER" at the top of the coin. It just so happened that Jason was visiting D.C. also. Through the wonderful world of social media, we realized we were there at the same time. We met up and I knew there would be no other person more deserving of my second coin than Sam's childhood best friend, Jason. I presented him the coin. Unbeknownst to him, I had another one in my pocket.

Years passed and Jason and I lost touch like so many people do. I thought I had completed a good portion of my book (before God changed my plans) and my editor told me she needed to learn more about Sam. It just so happened the night before I had a dream about Jason, and I was talking to him about Sam. I knew what I had to do. I got on social media and messaged him. Three weeks went by and I didn't hear anything. I wasn't shocked as I had not heard back from several of Sam's friends. Life is busy and many times, people don't reply to messages. I wasn't offended and decided to just wait.

I heard my phone chime. It was early morning and most of the world was still asleep. I looked down at my phone to see a message from

Jason. He had dreamt of Sam, woke up suddenly, and realized he had forgotten to reply to my message. Jason graciously and excitedly decided to be a part of this project. I knew he had a lot to offer and would be a wealth of knowledge regarding the past. There were some stories only a best friend knows, and that's what makes them so special.

After a couple weeks of correspondence through texting, Jason's tone turned serious.

"Renee, I have something I need to tell you." He continued, "The challenge coin you gave me. I lost it. I've torn my house a part looking for it. I even still have its indentation in my wallet. I can't find it. I believe someone may have taken it from me."

I couldn't help but have empathy for him. I explained that I forgave him and that sometimes those special things that have great meaning find their way back to us. I, too, have lost a coin. I've lost things, many things over the years, some with great personal value and miraculously they've returned to me. Some things were in the form of material possessions, some in the form of personal relationships.

"Well, Jason, I have a story to tell you. It may make you feel better."

"Back in 2010, Gentry graduated from Officer Training School in Montgomery, Alabama. Gentry had spent most of his military career as an enlisted member and worked his way through college and decided to get commissioned as an officer. Sam flew in for the weekend to see him get commissioned. Sam had encouraged him to take that journey, as difficult as it would be. Sam was never one to take the easy road, and expected no less from others," I said.

I continued, "What Gentry did not know at the time he was in OTS, was that tradition had it that after one gets commissioned, the first person to salute you should be presented with a silver dollar coin. Gentry called me the day I was traveling to Montgomery and said, 'I need a silver dollar coin.'"

"I stopped at every bank I came to and no one had a silver dollar coin. The only coin I could get was a gold dollar coin. Neither of us realized you had to order a coin from the Federal Mint. Gentry and I both realized we messed up. He had to present a gold coin with a promise to replace it with a silver dollar coin later."

I could feel the lump in my throat as I reminisced about this memory. "Gentry got commissioned and turned around. Chief Duvall was standing behind him. Chief was Gentry's supervisor for many years and came to see him get commissioned. She saluted him. *Oh God,* I thought. Gentry returned her salute and handed her the gold coin. He was embarrassed and promised her a silver dollar coin."

"Gentry felt humiliated and unprepared as a new officer. Sam, a new Major, approached Gentry, and saluted him. Sam reached into his pocket and presented Gentry with a beautiful silver dollar coin directly from the Federal Mint. The coin was sealed in a hard-shell plastic holder and placed in a blue velvet bag. He was so thoughtful and it was beautiful."

Tears filled my eyes, "Gentry eventually made it back to our home base and we hadn't had the time to order a new coin fast enough before he returned to work. He went into the office and Chief Duvall jokingly asked him about the coin. Again, Gentry was embarrassed. He had brought the coin Sam gave him to place at his desk. He reached into his pocket and gave her the coin that my brother had presented him. He had been a subordinate of hers for so long, he felt he had no other option and did not want to make her wait longer for a new coin. Even though he was now a Second Lieutenant, he still respected his former supervisors and revered them as leaders.

"Gentry and I were both sick over it, both hurt and a little angry at ourselves for relinquishing that coin. We only needed a little more time to order one for her, but it was too late. By the time ours came, there was no way we could have asked for it back.

"A year later, almost to the day, Sam was killed, and I immediately thought of that coin. It was the last gift Sam gave us and that coin flashed before my eyes. It was gone forever. I had no way of ever getting back the last gift my brother gave us. Now years later, I am still so grieved by this. My heart longs for the return of that coin my brother gave Gentry so many years ago on the most important day of Gentry's entire military career."

2010 - Gentry and Sam at Gentry's Officer Training School Commissioning, the last time he came to see us.
Courtesy of Renee Nickell.

Jason was quiet. He did not know what to say. He couldn't help but feel remorse over losing the coin, yet somehow, he must have felt a small amount of relief after I had lost such a treasure myself.

Suddenly, I remembered I had the exact replica of the coin I had given Jason many years before and I'd be seeing Jason in just two days. I called my mom and told her I was going to give him my coin. She said, "That's a wonderful idea. All you have to do is call the store in DC and they can ship you another one."

Our families met up for dinner and shared many stories about Sam. Jason told me how, at one point, Jason's bunk bed had collapsed and fell on Sam, and he had a nice goose egg to prove it. "You know what's weird, Jason?" I asked. "When I became older than my older brother…that was hard." Sam was 36 when he died and I'll never forget when I turned 36 as well. But when I turned 37, I realized I had, in fact, outlived my big brother.

Jason was always lightening the mood, "You know, rumor has it, that I have a birthday card from when I was born that has scribbling on it from Sammy. I presume he needed something to do while everyone was waiting for me to make my appearance."

We talked about having a bridge dedicated in Sam's honor, and my mission during this trip was to find the perfect bridge to have named after Sam. Jason suggested I should name the seven-mile bridge that goes to Key West. Sam loved going to Key West. My mom used to frequent The Keys with us when we were kids. We also remembered Sam's flight to be with his fraternity brothers while he was in training.

Jason told me about the time, probably in 2001, when Jason was in Nashville working. Sam had gotten Jason's exact location and told him an exact time to walk outside his building. Jason went outside just as Sam did an inverted flyover in his F18. "I could see his helmet,"

Jason said. "That's how close he was." The movie Top Gun came to mind and I wasn't surprised in the least. That was Sam. I'm sure he was laughing the entire time.

About the time we were about to say our goodbyes, I reached in my pocket and pulled out the extra challenge coin I had. I presented it to him. His face dropped. "This isn't your last one, is it?" he asked.

"Don't worry about it, I can get another." Sometimes it's worth giving up something valuable and of great meaning if it means bringing comfort to another person. Things are replaceable, people are not.

2017 - The coin I had engraved, but for some reason bought two that day.
Courtesy of Renee Nickell.

The next morning, we began the trek through the beautiful mountains of Ashville, North Carolina, on our way to our childhood hometown in the Raleigh area. *I could live here (in Ashville).* Gentry and I joked that if I became a best-selling author, he'd retire and we'd move to the mountains for the "simple life." He's worked at least two jobs since I met him. He has missed many anniversaries and birthdays serving our country. Now, as a disabled vet, we needed some tranquility in our life. I joked that I may only sell enough books to cover the gas from this little adventure we were on.

2017 - Renee at the entrance to Beaufort Air Station, SC.
Courtesy of Renee Nickell.

The beautiful, milky gray, Spanish moss draped from the trees over the road, eerily majestic, reminding me that we were, in fact, in the south. We were almost to Beaufort, South Carolina, where my brother had been stationed between 2001 and 2005. I was about to meet Ray, Noah, and Josh for the first time. When I decided to write this book, I knew I would need personal stories, but I never thought I would have an inside view of Sam's life during that time. The camaraderie, the friendships, and the bonds form a brotherhood between pilots. They're a different breed, shaped from a one of a kind mold. These men who knew Sam obviously had cared deeply for him. They were impacted greatly by his life…and his death. I was nervous for myself and for them. At no time did I desire for this journey to invoke more pain, albeit sometimes necessary discomfort to remember the past.

First, Ray and Noah met us at the Officer's Club, which happened to be closed that day. Ray made prior arrangements with the manager to allow us to see a large plaque that hung from the wall inside the "O Club." It was a large portrait of Saddam Hussein. It was quite obvious this picture had been hand crafted by a very talented artist. Surrounding the portrait were plaques with pilots' names from Operation Iraqi Freedom. I looked for Sam's name before I even addressed the significance of the portrait and the bullet hole sticker that covered Saddam's forehead, obviously placed as a spoof and a reminder of exactly why they were there. There it was…Capt Sam "Abbo" Griffith, his call sign given to him. After a flight, Sam would more often than not have an odor about him reminiscent of the Aborigines. "Abbo" stuck.

Noah began to tell us the significance of the painting. One of the crews had occupied a palace of Saddam Hussein. One of the pilots took out a knife and cut the portrait out of the frame on the wall to bring back as a gift for the 533 squadron. He had rolled it up and placed it in his bag until he returned to American soil. Months later, they opened it and had it framed, applying all the names of the pilots for that mission onto plaques. The bullet hole sticker was a nice, additional touch.

2017 - Portrait of Saddam Hussein hung in the Officer's Club, Beaufort Air Station.
Courtesy of Renee Nickell.

Noah and Ray led us to the back room where the smell of tobacco lingered in the air. This room was for a rowdier crowd, heavier drinking and smoking. Two 4'x 8' plywood planks were placed in the center of a brick wall due to years of excessive bottle breaking against the brick and mortar. With added stability to the wall, Marines could continue the tradition of breaking bottles. Broken glass still covered the floor. I was sure many stories of wild adventures had been shared between brothers within those walls.

Upon exiting, we headed to lunch. We went to a quaint little white, barn-type eatery. It could have come straight out of a magazine article for Town and Country. Beautiful hardwood floors ran the length of the inside. Mason jars with homemade jams and cookie mixes covered the shelves. As we enjoyed our lunch, the endearing stories began and I became quite aware in that moment, that this was the opportunity of a lifetime. I would have never taken this journey if I hadn't been writing this book.

A few stories were shared of Sam's piloting days. These guys obviously had the greatest job on earth, shocked that they actually got paid to do what they did as early twenty-somethings, beginning new families of their own. "You know, one thing I can tell you about Sam," Noah began. "He was so kind, a real genuine guy. I don't think I ever saw him have a bad day or be in a bad mood." He continued, "You know, we all have bad days, we all get down…but not Sam. He was able to brush things off and move past it without any bad attitude at all."

It was time to head to the squadron. We had an appointment to take a tour that Ray had coordinated for us. A tall, slender, pilot approached us at the secured gate to the aircraft maintenance and aircraft facilities. I noticed he looked strangely familiar, but I brushed it off that he must just have a familiar face. Noah leaned in and said, "Hey, Captain just got back from the 4th ANGLICO where he was the I&I (Inspector/Instructor) from 2015-2017."

"No kidding, I'm sure he must know Sam's Marines," I said.

We all entered the squadron where the Captain briefed us. He planned to show us around and then take us out to the flight line where we could get up close and personal with the F-18. He stated he had a slideshow to show us about Sam. It was a slideshow of his room dedication that took place in June of that same year at 4th ANGLICO in West Palm Beach.

"Wait, what?" I said. "You were there? I was there!"

He said, "You're his sister? The one who was at the 4th ANGLICO in West Palm for the ceremony with your mom?"

I smiled. We both were amused. My son and his son, both about the same age, bounced a ball all over the 4th ANGLICO just six months prior to this moment. He was the I&I who helped coordinate the room dedication. I remembered him well because he was unaffected nor concerned with my son's level of energy which I found a rare quality.

We walked into a back, large storage type room, covered in plaques and pictures of pilots. The room had been turned into a bar. On the wall was a large TV screen that was currently displaying a picture of Sam's plaque that hung outside his room at the 4th ANGLICO. Major said, "If you look closely, the tip of your and your mom's shoulders are there in the picture. I had cropped the picture just in case it was a different family member coming today."

In fact, it was me. I had that picture taken. I believe it was my phone that took that picture that was now on a TV screen nearly 500 miles away in a squadron my brother had served in over twelve years ago. I couldn't believe the irony.

I was standing there, quite uncomfortably I may add, as I now held the attention of about fifteen aircrew who were relaxing in their homemade bar, listening to Noah talk about Sam. I've never been good at receiving attention, and certainly the additional fact of the uncomfortable, yet honorable aspect that Sam was no longer with us, didn't help. They were all very kind and gracious and if I had been more of an extrovert, I'm sure I would have engaged myself more. This was already overwhelming for me. I couldn't believe I had this opportunity to meet people who had no idea who I was, just weeks prior, yet took time out of their day to make time for me and my family.

2017 - (left to right) Capt Oser, Noah, Josh, Renee, and Ray on the flight line of the 533rd.
Courtesy of Renee Nickell.

Capt Oser, Noah, Ray, and Josh took Gentry, the kids and me out to the flight line. There were about eight F-18's lined up and my children had the time of their lives, getting to see the inside of the cockpit. I looked up at my seven-year-old son, who was on a platform, ten feet in the air, beside the open cockpit. He stood there as serious as ever and saluted us. My heart melted. My children will always know and understand the sacrifice of our fallen. My son, who was only one year at the time of Sam's death, talks about Sam as if he knew him well. I am proud to say this is a testament to us as a family, as we continue to remember Sam and keep his memory alive. I am saddened for them, that they lost their beloved uncle in a war. Yet, I continue to be amazed at these children who love their country, our military, and understand more than many adults do.

We began to walk into the hanger doors as two F-18's landed. They were coming in for aircraft recovery. We needed to protect little ears. I turned around and looked out at the flight line filled with F-18's. The two jets' canopies opened and the pilots deplaned. I stood there, and imagined my brother, there at his squadron, coming in from a flight. There I was, years later, standing where he once served. He wasn't there and it wasn't Sam who had exited the aircraft. I felt the tears as they began to softly flow down my cheeks. I didn't want to cry. I didn't want anyone to see and I quickly placed my sunglasses on. I looked down and saw little Ethan. He was keenly aware that his mama was crying. He came and placed his little arms around my waist and buried his head in my waist. Leah, eleven years old, placed her arms around me as tightly as she could; neither said a word. Yet, they silently understood. It was time to leave…on to our next destination…my mom's home in Jupiter, Florida. We were about to spend a much-needed Christmas with her.

2017 - (left to right) Maggie, Gentry, Renee, little Ethan, Leah, and Kylee with an F-18, Beaufort Air Station, SC.
Courtesy of Renee Nickell.

Sentiments from Sam's Brothers

Over the course of the past six years, Mom and I have had the honor of getting to know several of the Marines that knew and loved Sam. I absolutely believe that not just I, but many reflect daily on how they can be better people especially when someone like Sam has impacted their lives in such a special way. The more I seek, the more healing comes my way, because I know that regardless of whether Sam is here on earth with us or not, his memory is alive, his legacy still impacts, and the relationship he had with each person was unique and meaningful. He was their mentor, their leader, and sometimes, like a father figure or an older brother to them.

One Marine in particular, Sergeant Anthony Cuesta, our family has come to know and love dearly. Anthony is a South Florida local who was attached to the 4th ANGLICO with Sam. I probably met Anthony at Sam's memorial service, but I have no recollection of meeting anyone, unfortunately. I would later meet him again at Sam's annual golf tournament held in Jupiter, Florida each June.

Anthony, a young, handsome Marine with a warm smile and welcoming disposition, was quite easy to talk with. He always spoke so fondly of Sam. I think part of his healing process in losing Sam was showing up to events that honored him, bringing celebration into a sorrowful situation. One thing I have learned is that while we all still mourn independently and sometimes together, people love to celebrate life. Grief is uncomfortable. There are no two ways around it. People would rather come to weddings than funerals, right? The same holds true with military traumas. We'd all much rather sit around and celebrate the life of a loved one, share a drink, and laugh until we cry. No one wants to sit around and cry. People have therapists for that. I get it. Sometimes, the tears happen and you can't help that, but there comes a time when the mourning can turn to celebration.

During one of these times, sharing drinks, and talking about Sam, Anthony began to share something very personal to him:

"I'll never forget the night I was awarded my Naval Parachutist Insignia ("Gold Wings") after fulfilling the required criteria. I had overheard someone once cite the fact that 3% of the Marine Corps are parachutists, and that 1% are awarded the naval parachutist insignia. I don't know if the numbers actually check out, but it surely is a small percentage."

He went on...

"What was most memorable about that night however, was neither the award ceremony, nor the fact that I was finally attaining what many in our unit strived for as a crowning achievement...what was most memorable was what Sam said to us that night. He told us to never take no for an answer and to be proactive about what we wanted in life. Don't wait for something to be done for you or expect things will just fall your way. If you don't make things happen, then things will happen to you.

"Sam knew how important getting our Gold Wings was to the jumpers in our detachment. He knew it was important to our morale, our sense of accomplishment, and our pride at ANGLICO. He did not take no for an answer when there was pushback about conducting parachuting night operations weeks before leaving for deployment. He relentlessly pushed for it to happen. He knew how important it was to us. Everyone knew that if Sam had not pushed for it, it never would have happened. This is the selflessness that true leadership is made of, the selfless nature of Sam. He did everything for others, never for himself. To this day, I recite a summarized version of that talk to myself in memory of him and that lesson. 'If you don't make things happen, then things will happen to you. Be proactive about what you want in life.' It's also a great opportunity to reminisce and honor the selflessness and true deep caring nature of a great man and leader I had the pleasure of serving under."

2009 - (Top left to right) George Mills, Sam, Errol Miranda,
(Bottom left to right) Jason Hartzell, Martin Castillo, Anthony Cuesta
training to go to Iraq at Camp Lejuene, NC.
Courtesy of Martin Castillo.

Staff Sergeant Ralph Perez had similar sentiments about Sam: "Sam had heard about me and heard I was decent. He took me under his wing when he was taking volunteers for the deployment. Sam came up to me and said, 'Can I count on you? Can I take a gamble on you?'

"And I said, 'Yes, of course.' This man had just met me and he made me a Team Chief after just getting promoted to Sergeant, and I was the only Sergeant Fire Power Control Team Chief on that deployment (to Afghanistan). I went up to him one on one. Even though I was happy about making Team Chief, I had concerns. I said to him, 'Sir, I have to be honest, I just got promoted to Sergeant and this is all new to me.'

"Sam said, 'I wouldn't have promoted you to Team Chief if I didn't think you were capable of handling it.' Sam squinted his eyes and gave me a little nod. At that point, I was panicked and scared because I knew I didn't want to let Sam down. Sam took me under his wing as a new Sergeant when I was a little rough around the edges. I knew my job, but I had to deal with knowing the next ranks job also. Sam helped me through the growing pains and he just always believed in me.

"Whenever I was kicking myself in the ass for making dumb mistakes, he never ever lost faith in me. Sam would do things for us just because he knew we wanted to achieve a goal. We talked about getting our Gold Wings right before deploying. Sam didn't have to, but he made calls. He got some crazy deals made with some pilots. He got us aircraft. He got us a base when we were in Virginia for training before deployment. It was one of the weekends we had off and he said, 'Hey, you guys want your Gold Wings? I got you pilots, I got you time.' There were five or six of us that got our Gold Wings. Sam didn't have to help us, but he did just because we wanted those wings so much. It was almost a liability because we could have gotten injured on a jump right before deployment and then we would have become undeployable.

"Sam cared so much about what we wanted that he made it happen. That's the type of guy he was. He would go any distance for his men and to this day, he was the best officer I ever worked for. He never gave up on me. Through my time with him, I went from being one of the new guys as a Sergeant to now being one of the guys that are looked up to in the company. I attribute that to him. To this day, I never want to let him down. I think about what Sam would do and I do that. That's who he was."

Staff Sergeant Errol Miranda was not only a devoted Marine to the 4th ANGLICO, serving alongside Sam, he was also Sam's friend. He added: “The first time I met Sam was at the end of his first drill at 4th ANGLICO. I knew we would make great friends when we made the same Supertroopers (movie) joke after the C.O. made a comment. The other thing that solidified our friendship was his ‘68 Mustang Fastback. The second he showed me the picture I said, ‘You made yourself a Bullitt!’

“He answered, ‘Damn straight.’

“When I got home from Afghanistan, to honor him, I purchased a 2008 Mustang Bullitt. I wanted a ‘68 but couldn't find one at a reasonable price.” He continued, “That picture he took in Africa with the radio handset was his least favorite picture of himself. Hartzell and I forced him to take it. Looking back, I am so glad we made him do it.”

2011 - Sam on training in Africa. He hated this picture, but it ended up being the most used photo of him in all the press.

As I was finishing up my manuscript, I received a message from one Marine, SSgt Castillo, who had served with Sam. I could feel his heart through his words which were incredibly moving. I could tell how much Sam meant to him. After expressing how sorry he was that they couldn't bring Sam home, he assured us that he would make sure Sam's memory was never forgotten. "We will tell stories to our kids about Sam, the hero that united us as a family, prepared us for a most difficult war, and got us all home to our families. We are forever in debt to the Griffith family."

He went on to write the following: "I had the opportunity to deploy twice with Sam. Iraq 2008-2009 and to Afghanistan 2011-2012. I had dealt with many officers during my time in the Marine Corps, but none were like Sam. Major G. was not your usual officer. He was a very intelligent officer who did not pretend that he knew it all or undermined anyone else. He knew some of us had already deployed and would ask our opinion and take our input into consideration. He would be hands on during training and not just leave it up to his Staff NCO's. He would make sure the I&I staff at the unit didn't mess up our pay or interfere with our training. Because of Sam, we had the proper tools and training for the mission to come.

"When we first arrived and started to separate into individual teams at Camp Leatherneck, we all said our goodbyes. Major G. said, 'Give me a hug, Castillo.'

"I said, 'Not till we're done. I'll see you in six months for the hug, sir!'

"Afghanistan was very difficult for all of us because we lost our leader. I remember being in my bed waiting on a Mission when Lt Debolt, now Lt Commander Debolt, called me over to the C.O.C. to break the news that we had lost Major G. that morning in a firefight. We tried to get to Bastion, Camp Leatherneck to receive Major G. and the Team, but birds were grounded due to the weather. I was really in disbelief and all I could think of was his two sons.

"After days passed and missions continued, my good friend Gunny Van Hof told me to do what Major G. would have wanted me to do. 'Lace up your boots and continue the good fight.'

"Remembering that Major G. prepared us for this fight kept me going. After being away six months, all the teams reunited at Camp Leatherneck. We hugged, we laughed, and then we had that awkward moment when we realized we were going home without our brother!

"Semper Fi, Brother, till we meet again." SSgt Castillo, Martin

2011 - Martin Castillo, Sam, and Ralph Perez during the Marine Corps Ball.
Courtesy of Martin Castillo.

I reflected on all their words about Sam. Time and time again, I heard the same thing, "Selfless, caring, leader, relentless, others before self." I, as Sam's sister, feel so humbled to have had such an honorable, caring man as my brother. Just when I felt like giving up writing this book because life happens and got hard, I pushed through. I had no other choice but to persevere, like Sam would have, like his Marines have…like he would have told me to do.

What has made this journey even more difficult is to know that his brothers in arms will always carry the weight and the burden of not bringing Sam home. I would never want that for them. I would never want them to feel that they somehow let our family down. Since Sam's death, they have done nothing but extend love and honor to our family. I know that their connection to us has been paramount in our healing. For that, we are forever grateful.

The Good Things

Through it all, we learned perseverance. The process began just two years after Sam was killed. In new grief, my mom decided she wanted to start an annual golf tournament as a charitable event to help other military organizations. The first year, my mom and stepdad rounded up many local friends, gathered donations for raffles and invited Sam's Marine friends. We were all crammed into a small golf clubhouse in Jupiter. I say this as a good thing. We had a great turnout. That first year in 2014, Mom and Donnie raised around $10,000 during the tournament which was held on Sam's birthday. By the fourth year, they had raised over $38,000 and became the largest charitable golf tournament in North Palm Beach County. The proceeds continue to benefit the Renewal Coalition of Jupiter.

"The Renewal Coalition is a non-profit whose sole mission is to assist wounded service members and their families in their transition from military to civilian life or back to active duty by offering a retreat at Florida residences with the intent of providing military families a relaxed, pleasant, and welcoming environment" (Renewal Coalition; n.d.).

During my visit to Mom's, we met with Mary Hinton at the Renewal Coalition property. As we drove through the shaded drive, we came upon tranquil and peaceful homes located directly on the Loxahatchee River. The water looked like glass as the sun beautifully reflected off its surface. The children immediately ran out onto the dock to search for any evidence of wildlife.

We sat with Mary. We talked about the property and the many families that are helped each year through the dedication, hard work, and love by the volunteers and employees of the Renewal Coalition. Many service members who return from Iraq and Afghanistan must go through a reintegration process, either because of combat related injuries, or simply back into their families. The Renewal Coalition helps to assist with that process.

2017 - (left to right) Mary Hinton, Renee, Mom
at the Renewal Coalition in Tequesta, Florida.
Courtesy of Renee Nickell.

"We focus on the family sharing time together without the worries of expenses. Lodging, meals, entertainment, ground travel and even the small amenities are the responsibility of Renewal Coalition and **NOT** the service members and their families" (Renewal Coalition; n.d.).

Since the year 2009, the Renewal Coalition has offered over 60 retreats, helped 368 wounded military personnel and 380 families of wounded military personnel (Renewal Coalition; n.d.). This is an incredible organization that helps to change the lives of many, many families each year.

Every year, Mom and Donnie work tirelessly to gain support from friends, family, and the community. Every year, generous sponsors donate graciously. Items range from small raffle prizes to large ticket items like golf outings to some of the best golf courses in Florida or tens of thousands of dollars of boating equipment. All these items are raffled off every year in memory of Sam.

The tournament has grown so much that we had to change venues. We now provide a catered breakfast; hors d'oeuvres, a full-service bar and dinner for all our supporters. This is a wonderful time for Sam's family, friends, and Marines to come together for the purpose of celebrating Sam's life and helping us all move forward. We get to share stories about him as well as a time to give back to so many who have sacrificed so much for our freedom. It is a time we all look forward to. The moment one golf tournament ends, we eagerly anticipate the next year when we can all come together again. It gives us purpose and meaning and helps us to somehow find some meaning in why Sam isn't with us anymore.

Some things we will never understand. All we can do is continue to move forward, forgive more, and love more.

2017 - (left to right) Jim Winters, Anthony Cuesta, Teddy Pernal, Renee, and mom at the Major Samuel Griffith Memorial Golf Tournament, Jupiter, FL. Teddy gave a very warm and endearing speech about Sam and 4th ANGLICO.
Courtesy of Renee Nickell.

A Time to Reflect

If your heart is available, God can use you, regardless of what you've been through or who you are. Believe in the purpose God has for you and your life. Jeremiah 1:5 reminds us that before we were born, God called us to a purpose. While the road has been bumpy with hills and valleys, and sometimes downright heartbreaking, I always knew I was made for more. I do not mean more than being a wife or a mother, because those are miracles in themselves. Made for more is to use the pain and the struggles we go through. It is to use the tiniest glimmer of hope we can hold on to, and fight with everything we have inside so that somewhere along the road, we can grab someone else's hand and inspire them.

If we can grasp the true nature of God's purpose, and trust that He will not lead us to failure, but will guide us, we can make it. My life's circumstances took my focus off the Lord many times. Like I said before, sometimes pain is the only way we learn, the only way our hearts are truly changed. I knew the kind of person I wanted to be. My brother's death taught me more things about myself than I could ever dream possible, but I had to allow that to happen. It isn't easy, but it can be done.

You can heal from the death of a sibling. You can find a new normal and a new place within the family even if it is much different than before. You can be proud of your sibling, still grieve, and have your own special identity. It doesn't have to be one or the other. You can have it both ways. I am a proud Gold Star sibling. My brother died in war. I am not the same person I was before he died. I am healed, but like missing a limb, I still experience pain. I still grieve but have found joy and happiness and meaning. I am also–me.

I reflect back to the year 2012. A few months after Sam died, I began training for the Marine Corps 10k in Washington D.C. I trained for the next nine months leading up to it. I ran and I ran, and I hated ever second of it, just like Sam hated running. I am NOT a runner, but I wanted to do this. I needed to do this. Come October the same year, my mom and I flew to D.C., both with the intent to run the race together. My mom was, in fact, a runner, so I had no hope of beating her time. My goal was simply to finish.

The morning of the race, the air was so cold. There was a mist that settled over the crowd. Thousands of people waited for the start. It was exhilarating.

As the race began, I set my pace. I lost track of Mom amidst the other runners. My earbuds were tucked in under my Semper Fi head wrap, and I was jamming to my favorite playlist, feeling on top of the world. About a mile in, I suddenly lost my footing. The tip of my toe slipped on loose gravel and I could feel my body propel forward. I knew I was going down and had no way of stopping the momentum.

After being airborne for a split second, I hit the ground. HARD. I laid there, in shock, wondering if I had broken or sprained anything. It seemed my whole body hurt and I couldn't pinpoint an exact location. Runners continued to run past me. I pushed myself up onto my knees and I heard someone screaming at me from the side. I looked up and it was a big, muscular Marine dressed in his BDU's. "Ma'am! Do you need assistance?" he yelled.

I felt like I was in boot camp. He stood there on the side waiting for a response. I looked up at him, not sure what I needed. "Ma'am, DO YOU NEED ASSISTANCE?" he yelled again.

I stared at him, embarrassed now that the shock was wearing off.

"Ma'am! If I assist you, you cannot finish the race! You will forfeit!"

Wait...I trained for nine months to forfeit in the first mile? Oh, hell no! I looked at him and nodded, "No."

Just then, an African-American woman slightly older than I, with a beautiful face and a kind smile, paused her race and approached me. She placed her arm around my arm and gently lifted me off the ground. I was hurt and in pain, and I knew I had over four more miles to run. She asked me my name.

"Renee. I'm really sore and I need to finish this race," I told her.

"Well, okay, Renee, I'm Michelle and you're going to finish this race then," she replied. She continued, "I'm going to run alongside you. I'm going to run with you and make sure you finish."

I replied, "I'm going to slow you down. I can't run faster than a slow jog."

She ignored my comment and we started jogging. She asked, "Who's that picture of on your shirt?"

"It's my brother, Sam. He was killed in action last December. I'm running for him," I told her.

"Well, then, we are going to finish the race together, for your brother, Sam."

Michelle, a complete stranger, ran with me. When my ankles were too sore to continue, she'd stop with me and then push me to go further. I wasn't sure I was going to make it, but she kept telling me what mile marker we were on down to the quarter mile. "A little bit further," she'd say. "Let's go, Renee, you got this. This is for Sam."

At the very end of the race, there is a steep uphill to the finish line. She stayed with me the entire time, until we crossed that finish line together.

I cried. She hugged me and I just cried. I missed him so much and I knew he would have been proud of me. Those are the moments you feel on top of the world, like a conqueror. I never saw her again. Perhaps she was an angel.

What I learned from that experience is that there are times when we fall and are hurting. We can lay there on the pavement and contemplate whether we can get up or not. We have a choice to forfeit or get back in the game. We may even ponder this for a while. Fear and pain can keep us from getting up and moving forward. Just when you think hope is lost, someone who is selfless, someone who cares more about finishing a race together than faster, will come along, and help pick you up and encourage you to the finish line.

I wanted to be that person, like Michelle was for me. I didn't know how to get there. So, I journeyed through hell. I faced demons I didn't want to face. Years went by and I kept on and kept on and kept on. I wanted to give up so many times. Along the way, there was always that one person who helped me get to the next step of healing. Sometimes it was my mom, a close friend, my husband, cousins or even strangers. It was a series of choices I had to make. When I overcame challenges, I received the confidence I needed to find value in myself.

God has shown me that despite the brokenness the pain of losing my brother has caused, I am capable of far more than I ever imagined. I wanted to quit so many times as I wrote this book because the pain was unbearable. I had to relive everything I wanted to forget. I had to pick myself up off that closet floor time and time again as those around me encouraged me to keep going. My brother died in war and that changed me, but it doesn't take away from who God calls me to be today. I chose to write this story to help other grieving siblings find their way.

I had no idea a little conversation in my friend's car would lead me to write and publish a book. I never dreamed I would actually finish even though I'd always talked about writing. Yet, here I am. As painful a process as this was, I knew it would honor Sam and give someone like you a glimmer of hope. I know Sam is shining down on me and is so proud of his little sis. I am proud of the woman, wife, mother, daughter, and friend that I have become.

I still miss my brother. I still grieve for Sam. I do. I miss him with everything inside of me. It is with the strength that his memory gives me and the continuing love for him that I step out and encourage you as well. When we choose to move through our pain and help someone else through theirs, we actually become healed ourselves.

Jack Canfield, author of *Chicken Soup for the Soul,* has said, "If only one person is changed by your story, it's worth it."

Maybe the one person that was changed by this story was me. To that I'd say, "It was worth it."

2012 - Renee after the Marine Corps 10K in D.C.
Courtesy of Renee Nickell.

RESOURCES

National Suicide Prevention Lifeline
1-800-273-8255
Available 24 hours, every day.
www.suicidepreventionlifeline.org

Paws for Reflection (Equine Assisted Therapy)
For information or to donate:
www.pawsforreflectionranch.org

The Fisher House Foundation
www.fisherhouse.org

Toys For Tots Foundation
www.toysfortots.org

Renewal Coalition
Assisting Wounded Service Members and Their Families
www.renewalcoalition.org

Tragedy Assistance Program for Survivors (TAPS)

www.taps.org

TAPS founder, Bonnie Carroll, began the organization in 1994 following the tragic death of her husband, Brigadier General Tom Carroll, who was killed in an Army C-12 plane crash in 1992. Bonnie saw a need for military families to gain support through other peers who had also experienced the same tragic loss (Tragedy Assistance Program for Survivors; 2018).

Since 1994, TAPS has helped over 70,000 family members. "There are literally over 70,000 people who are not going to go through their grief alone," stated TAPS staff-member Kristen Buergey. Think about this for a moment. There were over 7,000 soldiers killed in both the Iraq and Afghanistan war since 2001. This does not include those who died from long-term combat related injuries, PTSD, suicide, or accidents. "TAPS is truly the family you never wanted. It's the elite club that you never want to make it into," she continued.

The core of TAPS has been the countless peers who volunteer their time to support parents, widows, siblings, and children who have experienced the unique loss of a military death. TAPS has provided support through many facets that offer life-saving tools for families that do not have any connections in the military. Many family members battle depression and anxiety from the trauma of their loved one's death. TAPS brings in some of the best professionals in the country in different areas of grief expertise to help families make connections, process grief, and move forward through their journey. Hope becomes tangible when you have none.

TAPS is well known for their hugs, care, and compassion, regardless of circumstance. It's a place of safety and refuge and reminds the grieving they are not alone and are not forgotten. TAPS sends parents, spouses, children, and siblings a grief box for every military related death that occurs while serving on active duty. That's a lot of love that goes out to these families.

I know there were times in my own grief walk when I literally felt I was going crazy. I was so numb, it was as if I couldn't even allow myself to feel the pain. I was simply existing. This is where TAPS steps in and holds your hand. They love you through this period by offering compassion. They taught me to have compassion towards myself.

There are many areas of compassion care that TAPS can offer to families: Peer support, seminars (regionally and nationally), retreats and expeditions. There are supports for every step of the way. Some of these opportunities make a way for families to not only grieve together, but honor their loved ones, provide an outlet where they feel safe and have fun too. They have provided the means for families to attend NFL games, meet the players, and honor their loved ones on the field during the game. They have taken young adult Gold Star children to sky dive with the Army Parachute Team, the Golden Knights. Families have been invited to the White House for special events and to participate in the annual Easter Egg hunt.

Travel opportunities include men, women, sibling, and widow retreats all over the country. The cost is only $50 for the family member to attend, plus transportation to and from. Expeditions like hiking Mount Kilimanjaro or Machu Piccu encourage family members to power through their grief and other obstacles in their lives. It also encourages them through the use of peer support to help them overcome mental, emotional, and physical challenges.

TAPS has a vast amount of support for suicide survivors. It has helped bring healing to thousands of families. A particular area close to my heart is that of sibling loss. We are considered "the forgotten mourners". TAPS recognizes the sibling as a valid part of the structure of a family. While most of the time, siblings feel like they are the last ones to be validated, TAPS reversed this and validates them in their grief along with every other family member. Siblings have given TAPS feedback on what they have needed and TAPS has accommodated those needs. There is a tremendously large peer support for siblings where they are able to interact and support one another.

If you or a loved one are in need of grief assistance through your grief-walk, TAPS contact information will be listed in the appendix of this book. Bonnie Carroll has done a tremendous work to make sure no one is forgotten. So much so, she received the 2015 Presidential Medal of Freedom, and continues today to make a difference and give hope to those that have none. She is one of the kindest, most compassionate people I've ever had the honor of meeting. She has seen every situation possible within families and continues to love them through the messiness of life and death.

2018 - (left to right) Renee, Kristen Buergey, and Heather Stang. Regional TAPS Seminar at Fort Hood, Texas. Heather is an Author, Speaker, and Coach that teaches mindfulness during grief.
Courtesy of Renee Nickell.

ABOUT THE AUTHOR

Author Renee Nickell loves all things patriotic, military, and supports military families. As a Gold Star sister and military spouse to her darling husband Gentry, she is very familiar with the pride and pain that flows along with service. Renee penned her first non-fiction memoir, *Always My Hero*, based on the life and tragic death of her beloved, slightly older and wiser brother, United States Marine, Major Samuel M. Griffith. In 2011, Sam was killed in action in Afghanistan which ultimately changed the course of Renee's life.

Renee now advocates for other Gold Star families. Her passion is to increase awareness of what the whole family left behind goes through so that we can better help them endure and recover. She also works with her mom and stepdad who hold an annual golf tournament in Sam's memory to raise support for the Renewal Coalition and service members wounded in combat.

She has learned to appreciate those things that help the heart heal and stories that inspire. A quality she and her brother shared, and a prerequisite for any of her close friends, is her love of laughter–for it truly is the best medicine.

Renee is an active member of her church in Fort Worth, Texas and has served in youth ministry, and small group studies in her home. Renee and her husband have four children, three girls and one boy, as well as three male dogs to help her husband and son even out the playing field.

For Speaking Requests and to sign up for her newsletter, visit
www.ReneeNickell.com
Follow Renee on Social Media:
Instagram: @renee_nickell
Facebook Official Page: @rmnickell
Twitter: @rmnickell

WORKS CITED

The Bible. Authorized English Standard Version. Good News Publishers, (n.d.).

Canfield, Jack. (2018). *Bestseller Blueprint*. www.bestsellerblueprintsystems.com

Carroll, Bonnie, and Alan D. Wolfelt, PhD. *Healing Your Grieving Heart After A Military Death*. Prod. Companion Press, 2015.

Hillsong United. "Oceans (Where Feet May Fail)". *Zion,* Hillsong Music and Sparrow Records, 2013.

Hillsong United. "When I Lost My Heart to You". *Empires*, Hillsong Music, Sparrow Records and Capitol Christian Music Group, 2015.

Renewal Coalition. Retrieved from: http://renewalcoalition.org (accessed 2018).

The Coral Snake. Retrieved from: http://www.thecoralsnake.com/griffith2 (accessed 2017).

Tragedy Assistance Program for Survivors. Retrieved from: https://www.taps.org/ (accessed 2018).

CPSIA information can be obtained
at www.ICGtesting.com
Printed in the USA
BVHW010046071118
532411BV00002B/3/P